Population and Community Biology

PREDATION

Population and Community Biology

Series Editors

M. B. Usher
Senior Lecturer, University of York, UK

M. L. Rosenzweig
Professor, Department of Ecology and Evolutionary Biology, University of Arizona, USA

The study of both populations and communities is central to the science of ecology. This series of books will explore many facets of population biology and the processes that determine the structure and dynamics of communities. Although individual authors and editors have freedom to develop their subjects in their own way, these books will all be scientifically rigorous and often utilize a quantitative approach to analysing population and community phenomena.

PREDATION

Robert J. Taylor

Associate Professor of Zoology
Clemson University
USA

NEW YORK   LONDON

CHAPMAN AND HALL

First published 1984 by
Chapman and Hall Ltd
11 New Fetter Lane, London EC4P 4EE
Published in the USA by
Chapman and Hall
733 Third Avenue, New York NY 10017

© *1984 Robert J. Taylor*

Printed in Great Britain by
J. W. Arrowsmith Ltd., Bristol

ISBN 0 412 25060 8 (cased)
ISBN 0 412 26120 0 (paperback)

British Library Cataloguing in Publication Data

Taylor, R. J. (Robert John)
    Predation. — (Population and community biology)
    1. Predation (Biology)
    I. Title   II. Series
    591.53   QL758

    ISBN 0-412-25060-8
    ISBN 0-412-26120-0 Pbk

Library of Congress Cataloging in Publication Data

Taylor, Robert J., 1945–
    Predation

    (Population and community biology)
    Bibliography: p.
    Includes index.
    1. Predation (Biology) I. Title. II. Series.
    QL758.T38   1984   574.5'3   84–4974
    ISBN 0-412-25060-8
    ISBN 0-412-26120-0 (pbk.)

Contents

Preface

When assuming the task of preparing a book such as this, one inevitably wonders why anyone would want to read it. I have always sympathized with Charles Elton's trenchant observation in his 1927 book that 'we have to face the fact that while ecological work is fascinating to do, it is unbearably dull to read about . . .' And yet several good reasons do exist for producing a small volume on predation. The subject is interesting in its own right; no ecologist can deny that predation is one of the basic processes in the natural world. And the logical roots for much currently published reasoning about predation are remarkably well hidden; if one must do research on the subject, it helps not to be forced to start from first principles. A student facing predator–prey interactions for the first time is confronted with an amazingly diverse and sometimes inaccessible literature, with a ratio of wheat to chaff not exceeding 1:5. A guide to the perplexed in this field does not exist at present, and I hope the book will serve that function.

But apart from these more-or-less academic reasons for writing the book, I am forced to it by my conviction that predators are important in the ecological scheme. They play a critical role in the biological control of insects and other pests and are therefore of immediate economic concern. They make up a disproportionate share of the world's threatened and endangered species and as such are of long-term ecological concern. They play a major and increasingly recognized role in structuring biotic communities. Yet management policies sometimes exhibit remarkably little awareness of the fundamental nature of the predator–prey interaction. For example, it is typical for predator and prey in a commercial fishery each to be harvested separately according to guidelines which take no account of their interdependence.

My attempts here to fill the niche I have just described must inevitably reflect countless conversations about predation with teachers, colleagues, and students at the University of California at Santa Barbara, Princeton University, The University of Minnesota, and Clemson University. For advice on various portions of the book I am indebted to P. K. Dayton, D. G. Heckel, P. A. Jordan, R. O. Peterson, M. L. Rosenzweig and M. B. Usher. In addition, I am especially grateful to the many fine graduate students with whom I have had the privilege of associating while teaching courses on predation at Minnesota and Clemson.

I dedicate this volume to my wife, Susana.

Clemson, South Carolina

Acknowledgements

Acknowledgements are made to the following for permission to reproduce or modify tables and figures: figure 2.1, from the *South African Journal of Wildlife Research*; figure 2.2, from the *Journal of Wildlife Management* (Copyright by The Wildlife Society); figure 2.3, from *Ecological Modelling* by the authors and the Elsevier Scientific Publishing Co.; figure 2.4, G. C. Haber; figure 4.1, R. O. Peterson; figures 4.2, 7.4, from *Hilgardia* by the Division of Agricultural Sciences, University of California; figure 4.3, J. H. Connell; figures 4.4, 4.5, from 'The importance of predation by crabs and fishes on benthic infauna in Chesapeake Bay' by R. W. Virnstein, *Ecology* (1977) **58**, 1199–1217 (Copyright © 1977 by the Ecological Society of America); figures 4.6, 10.4, from *Wildlife Monographs* (Copyright by The Wildlife Society); figures 5.2, 5.7, 6.1, 10.3, from the *Journal of Animal Ecology* by Blackwell Scientific Publications Ltd.; figures 5.4, 5.5, 10.1, from the *Journal of Bacteriology* by the authors and the American Society for Microbiology; figures 5.6, 6.4, 6.5, 6.6, 7.8, 7.9, by *Researches on Population Ecology*; figure 6.7, from the *Canadian Journal of Zoology* by the National Research Council of Canada; figures 7.5, 7.6, 7.7, from a paper by D. Pimentel, W. P. Nagel, and J. L. Madden in *The American Naturalist* by the University of Chicago (Copyright 1963 by the University of Chicago); figure 8.6, from *Ecological Modelling* by the author and the Elsevier Scientific Publishing Co.; figures 8.8, 9.4, 11.5, from the *Canadian Entomologist* by the Entomological Society of Canada; figure 9.1, from a paper by B. L. Partridge and T. J. Pitcher in *Nature* **279**, 418–419 by the author and Macmillan Journals Ltd (Copyright © 1979 by Macmillan Journals Ltd.); figure 9.3, from *Oecologia* by Springer-Verlag; figure 11.1, from a paper by S. A. Levin and J. D. Udovic in *The American Naturalist* by the University of Chicago (Copyright 1977 by the University of Chicago); figure 11.2, from *Fermentation Advances*, D. L. Perlman (ed.) by the authors and Academic Press; figure 11.3, from a paper by M. T. Horne in *Science* (1970) **168**, 992–993 by the author and the American Association for the Advancement of Science (Copyright 1970 by the American Association for the Advancement of Science); figure 11.4, from 'A complex community in a simple habitat: an experimental study with bacteria and phage' by L. Chao, B. R. Levin and F. M. Stewart, *Ecology* (1977) **58**, 369–378 (Copyright © 1977 by the Ecological Society of America); figures 11.6, 12.2, 12.3 from the *Cold Spring Harbor Symposium on Quantitative Biology* **22**, 139–151 by the author and the Cold Spring Harbor Laboratory (Copyright 1957 by the Cold Spring Harbor

Laboratory); table 4.3, from 'A predator–prey system in the marine intertidal region. I. *Balanus glandula* and several predatory species of *Thais*' by J. H. Connell, *Ecological Monographs* (1970) **40**, 49–78 (Copyright © 1970 by the Ecological Society of America).

1 Predators and predation

If there are any marks at all of special design
in creation, one of the things most evidently
designed is that a large proportion of all
animals should pass their existence in
tormenting and devouring other animals.

J. S. Mill (1874)

The first task faced by the writer of any book is the setting of limits around the
material to be covered. This is primarily an exercise in definition and, as with
any such exercise, is an invitation to dispute. Nearly every textbook dealing
with ecology offers a unique definition of predation. The lack of consensus
stems, I suggest, from variation in the extent to which we recognize that,
whatever it may be ecologically, predation is a clear, unambiguous behavioral
act. Other interactions of species, competition or mutualism for example,
imply no particular behaviors, and their definitions emerge solely from their
ecological consequences. By comparison, predation may be defined as a
behavioral act, an ecological process, or more commonly as some combination of both.

A secondary but nonetheless important difficulty with definition is that the
predatory habit is correlated with taxonomic classification. The mention of
predatory birds brings to mind the Falconiformes and Strigiformes, of
predatory mammals, the Carnivora. With the exception of parasitism, other
ecological interactions bear no association with taxonomy. While most
academics would be surprised and amused to hear a student propose to study
the competitive insects, a treatment of insect parasitoids would be deemed
perfectly appropriate, the group being limited to certain wasps and flies.

The interlacing of these influences tends to produce an image in our minds of
the archetypal predator as a fierce animal, well equipped for killing. However
we modify this species as adults, the type specimen remains the stuff of
childhood nightmares. Our attempts to generate rational definitions are
inevitably clouded by such mental images. I illustrate the resultant dilemma by
looking briefly at a fascinating predator, the extinct saber-toothed cat,
Smilodon californicus.[1]

Smilodon is undoubtedly one of the most impressive carnivores ever to have
graced the mammalian phylogenetic tree. The largest of the machairodont

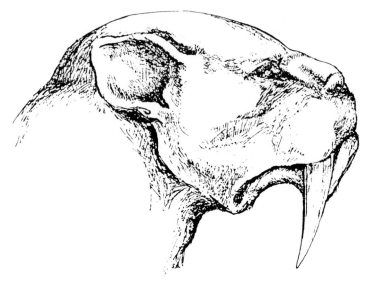

Figure 1.1   A reconstruction of the head of *Smilodon californicus* (from Miller, 1969).

cats, its hind limbs were comparable in size to those of the largest living felines; the fore limbs, by comparison, were enormously well developed. The massive and powerful neck and head wielded the most striking feature of this cat, the large, dagger-like canines (Fig. 1.1) which were used to stab its prey, the large herbivorous mammals of the late Pleistocene (Simpson, 1941).

Merriam and Stock (1932) published an interesting analysis of the morphology of *Smilodon*, in which they noted a couple of curious features of its skull; the nares were placed well back on the head compared to modern cats, and the palate contained a well-developed pair of grooves bounded by ridges which ran from the canines backwards to the throat. The combination of these features with the massive forequarters suggested to Merriam and Stock that *Smilodon* may have gained some portion of its nutrition from blood sucking. The sabers could have pierced the tough hide of the prey and served, with the forepaws, to anchor the cat while it drank its fill. The hair of the prey would not have blocked the nostrils. The benefits of this sort of strategy are clear; a wounded animal may heal, to be attacked again later. For all its size, *Smilodon* could not be expected to consume more than a small fraction of a dead mastodon or giant ground sloth before scavengers and spoilage destroyed the rest.

That a carnivore as large as *Smilodon* could live on blood alone is unlikely, given the concentration of protein in such a food. Its dentition, moreover, was clearly suited to eating meat. The carnassials were large and well designed for cutting; the incisors of fossil specimens show signs of wear consistent with the habit of worrying meat free from a carcass (Miller, 1969). Nonetheless, even the possibility that one of the most spectacular carnivores in the history of the Mammalia could have found its niche as an ectoparasite gives cause for reflection on the reliability of our biases about what predators should look like.

The elimination of morphological bias is unfortunately not the most difficult step toward a definition of this class of ecological interactions. The most difficult step is deciding how specific such a definition should be or, the other side of that coin, how many exceptions one can tolerate. The following sequence of definitions, gleaned both from current textbooks in ecology and from common usage, rank approximately in ascending order of generality.

(a) Predation occurs when one organism kills another for food

This is the most common definition. Some would restrict it to animals as predator and prey; others would allow plants on either end of the act. The identifying feature of this definition is that it requires both a behavior and the death of the prey. Its general form includes the consumption of seeds and some types of grazing; parasitism, however, is excluded.

(b) Predation occurs when individuals of one species eat living individuals of another

While not yet catholic, this definition embraces all of definition (a) plus herbivory and all types of parasitism. The difference stems from the relaxation of the requirement that the prey must die. Preference for this definition frequently arises from recognition of the arbitrary division between parasitism and predation. An adult insect parasitoid behaves like a predator; it searches, pursues and attacks, although the inevitable death of its host is caused by the larval offspring. The ecological impact of this particularly virulent disease organism differs only in detail from that of the more familiar carnivore and ought, in the minds of many, to qualify parasitoids as predators.

(c) Predation is a process by which one population benefits at the expense of another

This definition differs from the first two in taking absolutely no cognizance of the types of behavior involved. It focuses only upon the ecological results of the process. Mathematical theoreticians tend to favor this definition; it amounts to a verbal description of the form of their equations.[2] One could also argue against this definition on the grounds that it includes, under the rubric of predation, such phenomena as Batesian mimicry. But the primary difficulty with a strictly ecological definition is that the conflict with our behavioral image of the act can be severe. Valerio (1975) provides a nice example of this. A small scelionid wasp, a species of the genus *Baeus*, parasitizes egg clusters of the spider, *Achaearanea tepidariorum*. Survival through the first instar of the hatchling spiders is quite low in unparasitized egg masses. Apparently the spiderlings tend to sit around until they starve. The activity of emerging *Baeus* in the parasitized egg masses stirs the healthy spiderlings into cannibalism of their weaker siblings. The result is an increase in the percentage surviving to the second instar. By definition (b) the exploitation of spider eggs by the wasp constitutes predation; by definition (c) it does not.

(d) Predation is any ecological process in which energy and matter flow from one species to another

This definition has the advantage of no exceptions; nothing that even smacks of predation is excluded. But its inclusion of all heterotrophic organisms amounts to an abrogation of the task of expediting communication among scientists. Even theoreticians bridle at the inclusion of decomposers; such creatures require different sorts of equations.

This is not an exhaustive list. Which of these, or any other, definitions is the proper one remains largely a question of utility. An ecological behaviorist will consider (a) the most reasonable; an ecological philosopher might settle upon (d). For purposes of research I prefer (b), but this is largely a personal preference and reflects my bias that predation is more properly defined by the act rather than by its consequences. Its consequences, as we shall see, are not always obvious.

In writing this book, I chose to adopt definition (a), for the most part restricting it even further to include only animals killing animals. Insect parasitism is treated to some extent, inasmuch as its literature has profoundly influenced ecological theory. I offer no defense of this decision except the utilitarian one that it serves to bite off a barely manageable chunk of the literature.

NOTES

1. I shall normally refer to all species by their common names, unless the common name is either less well known or does not exist. Scientific equivalents are listed in the Appendix. Incidentally, comments such as this are placed in footnotes because they are, indeed, peripheral to the text. You may eventually find such notes tiresome if you insist upon reading each one at first encounter, a practice I strongly discourage.
2. Depending upon the purposes of the theoretician, definition (c) can be recast. For an examination of the coevolution of predator and prey, we might use the following:

   A predator population exploits a victim's population if and only if (i) the

victim's mean fitness is depressed by increase of its predator, and (ii) the predator's mean fitness is enhanced by increase of its victim.
Rosenzweig (1973b)

Or if we wanted to do a food web analysis with graph theory, we could use this one:

If two species are involved in a 2-cycle, and if the 2-cycle involves one '+' line and one '−' line, then the species may be regarded as predator and prey, or parasite and host.
Jeffries (1974)

2 Predation theory

And as imagination bodies forth
The form of things unknown, the poet's pen
Turns them to shapes, and gives to airy nothing
A local habitation and a name.

Shakespeare (*A Midsummer Night's Dream*)

The subject of this book is the influence of predation upon the dynamics of prey and predators; this chapter and the following form a conceptual foundation for investigating that influence. The majority of general ideas about predation being discussed today are mathematical in form, a fact which many ecologists openly deplore. We spend the greater part of our professional lives impressing one another with the complexity of the natural world and find it irritating to have an impertinent theoretician reduce that complexity to a simple set of equations. Nonetheless, I maintain we should welcome such a trend towards mathematical theory as a necessary manifestation of the growing maturity of ecology as a discipline. The development of theory of any sort requires, as a necessary first step, abstraction and simplification of the processes under investigation. Given the complexity of predation, the theory of this process will best emerge from precisely defined assumptions manipulated with impeccable logic.

An unfortunate but perhaps inevitable side effect of the transition from verbal to mathematical models has been the partitioning of students of predation into strict theoreticians and strict empiricists. A single scientist cannot do everything well, and the increasingly sophisticated uses of mathematics in ecology take time and a lot of effort to understand. As the partition grows higher each year, fewer and fewer of us find fence-sitting either comfortable or respected. Assuming that good fences make bad scientists, I shall attempt in this and the next chapter to put into digestible form some of the more robust conceptual results of mathematical predator–prey theory. The strictly field-oriented reader will be tempted to skip portions of this chapter, a predilection with which I confess some sympathy. I urge you to fight the temptation and work through the chapter carefully. It contains an absolute minimum of mathematics, and the perspective it provides underpins the remainder of the book.

The chapter opens with a discussion of three studies in which simple mathematical theory has been usefully employed in the analysis of real ecological problems involving predators. Following this discussion is a short exposition of a basic, general model, which I provide not as a single conceptual framework from which specific models can be derived but rather as a basis for examining both the basic procedures used and some of the fundamental issues that arise in theorizing about predatory interactions.

The case histories chosen represent neither the acme of the modeler's craft nor the most elegant applications of theory to the real world of ecologists. They resemble one another in their subject – the management of big game populations subject to predation, and in their general mathematical form – difference equations used in computer simulation. Beyond this, they incorporate predation in quite different ways and are instructive in that respect.

The first example treats the interaction of ungulates and large carnivores in South Africa. During seven of the nine years between 1961 and 1969 the Central District of the Kruger National Park received less than the average amount of rainfall, culminating in 1969–1970 with the worst drought in 34 years. Populations of zebra and wildebeest had either increased or remained constant during this period, and the range, as a consequence, began to deteriorate. Officials of the National Park decided to forestall a catastrophic decline in the numbers of grazing herbivores by cropping the herds of zebra and wildebeest until the drought ended and the carrying capacity of the rangeland returned to acceptable levels. The weather improved in 1970, the grasslands rebounded quickly, and the cropping program was phased out gradually, terminating in 1972. As expected, the wildebeest population began to decline in 1969 and continued to drop through 1970 and 1971. Contrary to expectations the wildebeest herd continued to shrink after 1971, despite both relaxation of the harvest and improved habitat. By 1975 it had declined to less than half its size six years earlier. The questions facing the Park's game managers were why the planned reduction was followed by a further unplanned decrease and what role did the Park's large lion population play.

Starfield, Shiell and Smuts (1976) constructed a simple model of the wildebeest population to address this question. The model consisted of a set of four finite-difference equations, one each for the four age classes into which the wildebeest population was divided. The numbers represent animals alive at the end of the calving season.

$$C_{i+1} = \alpha W_{i+1} + \alpha' T_{i+1} \qquad (2.1)$$

$$Y_{i+1} = \beta C_i \qquad (2.2)$$

$$T_{i+1} = Y_i - \frac{\gamma L_i Y_i}{Y_i + T_i + W_i} \qquad (2.3)$$

$$W_{i+1} = W_i + T_i - \frac{\gamma L_i (T_i + W_i)}{Y_i + T_i + W_i} \qquad (2.4)$$

The variables and constants are defined as follows:

C_i = the number of calves, new-born
Y_i = the number of yearling animals
T_i = the number of two-year-old wildebeest
W_i = the number of adults older than two years
L_i = the number of lions
α = the births per adult wildebeest = 0.45/year
α' = the births per two-year-old = 0.15/year
β = the proportion of calves surviving the first year
γ = the number of kills per lion per year

The constants L_i, β and γ were allowed to vary from year to year. The builders of this model chose to exclude any factor not specifically known to influence the wildebeest population between 1969 and 1975. In particular, because the primary question was why wildebeests declined in the face of improved habitat quality, the model excludes feedback terms which would allow the population to limit its own numbers by intraspecific competition. Note in this respect that the constants α, α', and β, relating to reproduction and survival, do not depend upon the numbers of wildebeest. The only density feedback comes in the predation terms, the second terms in the third and fourth equations. For several reasons the model contains no equation for changes in the numbers of lions; their numbers changed little between 1969 and 1975 (from a low of 500 to a high of 550); the authors were not sure of the mechanisms governing fluctuations in lion populations; and wildebeests formed only 28% of recorded lion kills between 1969 and 1975.[1] In the simulation of Equations (2.1)–(2.4), the numbers of lions were varied from year to year to reflect census estimates. The style of predation was quite simple; lions were assumed to remove yearling, two-year-old and adult wildebeest in proportion to their numbers in the population.

Beginning with the wildebeest population of 1969, the model generated the population trajectory in Fig. 2.1, indicating quite clearly that lion predation was adequate to explain the continued decline of the herd after the drought had broken. Equations (2.1)–(2.4) also allowed Starfield, Shiell, and Smuts to explore what might have happened if the numbers of lions had been reduced to 450. The dashed line in Fig. 2.1 suggests that such a policy might have minimized the drop in wildebeest numbers.

The goals of the modellers in this study were modest: they sought only to account for a short-lived and unexpected phenomenon. To this end this model was appropriate even though Equations (2.1)–(2.4) are extraordinarily simple for a management model. Its parameters are sufficiently few that these could be estimated, even from the inadequate data available at that time.

The second example treats the management of populations of moose in southwestern Quebec. The proximate goal of this theoretical effort by Crête *et al.* (1981a,b) was to explain the results of an experimental manipulation of

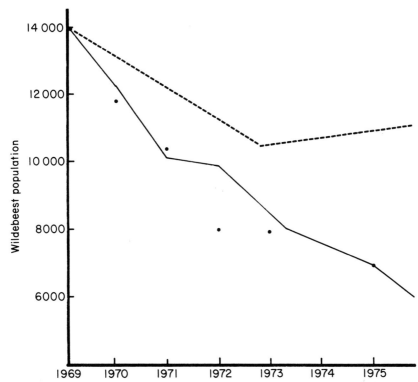

Figure 2.1 A comparison of census records of wildebeests in the Kruger National Park (dots) to predictions of the model (solid line). The dashed line represents the model's predictions of what would have happened to wildebeest numbers had lion densities been reduced (after Starfield *et al.*, 1976).

hunting pressures. The ultimate goal was to discover how this ungulate can be managed for maximum hunting success in the face of predation by wolves. Figure 2.2 shows the study area. Each of three control areas in which normal hunting pressures existed matched an experimental area of comparable vegetation and topography in which hunting was reduced to about 5% of the level in the control areas. Census data suggested that moose numbers were fairly stable in both treatments, that moose were not limited by the food supply, and that the principal cause of natural mortality was predation by wolves.

The authors of this study asked a great deal more of their model than did Starfield *et al.*, and their model is correspondingly more complicated. They chose to use a discrete-time model, primarily to reflect the highly seasonal processes of reproduction, predation and hunting. The moose population was subdivided into 20 year-classes for which different survival and reproduction parameters were estimated. Each age-class, in turn, was subdivided into a

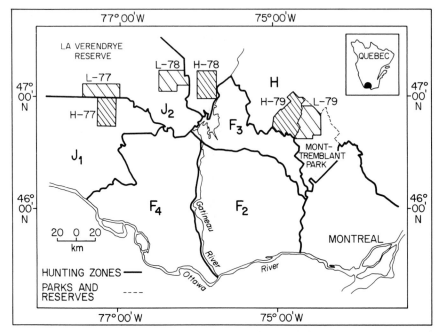

Figure 2.2   Map of the study areas in Quebec. Blocks labelled H represent areas of high hunting pressure; those labelled L had low hunting pressure. The numbers are the year in which the block was surveyed. The areas F_2, F_3, F_4, H, J_1, and J_2 are different hunting zones (from Crête *et al.*, 1981b).

density at the end of the summer and a density at the end of the winter. If $M_{fwi}(t)$ and $M_{mwi}(t)$ are the numbers of female and male moose at the end of the winter of age i in year t, then

$$M_{fw1}(t) = M_{mw1}(t) = 0.5 \sum_{i=2}^{20} b_i M_{fwi}(t) \qquad (2.5)$$

The parameter b_i is the fraction of cows of age i bearing calves. Of these newborn a number are lost by predation in their first summer; the survivors number

$$M_{fsi}(t+1) = M_{fwi}(t+1) = M_{fwi}(t)\, e^{-aW(t)} \qquad (2.6)$$

in which $W(t)$ is the density of wolves in the summer of year t and a is the rate of killing per wolf. The surviving young and adult moose then run the gauntlet of the hunting season, which was assumed to be the only mortality source other than wolf predation. Survival over the winter was of the same form as Equation (2.6), except that the rate of predation was made age-dependent; old and young moose were more vulnerable than prime adults.

Each kill was multiplied by the average weight of a moose of that age to compute the total amount of meat available to wolves over the winter. If this were sufficient to maintain that winter's wolf population, no wolves starved;

wolves in excess of the sustainable number were assumed to either leave or starve. And unlike the moose, the wolf population was not subdivided by age. Three alternative models of the life history of the predator were employed: (i) in the first model the spring production of pups per wolf was assumed to be constant; (ii) in the second the reproductive rate was related directly to winter survival, presumably reflecting the nutritional status of the pack; (iii) in the third model the reproductive rate varied as in model two, but survival was improved by the presence of garbage dumps as alternative food sources when moose were uncommon.

The three graphs in Fig. 2.3 reveal the results of the different assumptions. Figure 2.3(a) derived from the first assumption, that wolf reproduction is constant per adult. The result was a cycle of about 20 years in both populations. The imposition of moose hunting seemed to increase the amplitude of oscillations. Figure 2.3(b) explores the influence of a wolf reproductive rate sensitive to winter survival. This element of negative feedback is apparently sufficient over a very long period of time to stabilize the population, but that stability is lost when hunting is allowed. Census records suggest that hunting in Quebec has not destabilized moose populations, so neither models 1 nor 2 were of interest to the authors. The simulation of model 3 generated the population trajectories in Fig. 2.3(c). Apparently the addition of garbage dumps to model 2 stabilizes this system. Of equal importance to the authors, that stability did not disappear with the addition of hunting.

These are reasonably simple equations, apart from the detailed consideration of the age-structure of the moose, yet their third variant generates qualitatively adequate predictions. The model has the additional important attribute that it suggests a simple experiment: fence the garbage dumps and look for a moose population which is larger and, on average, more prone to oscillation.

Gordon Haber's work on the big game species of Denali National Park in Alaska provides the third example (Haber *et al.*, 1976; Haber, 1977). His field studies culminated in the development of a large simulation model of the interaction of a typical wolf pack with its herds of moose, Dall sheep, and woodland caribou. The model is much too bulky to present here; its form is a system of stochastic difference equations grouped into four subsets. The first describes the severity of winter weather, a critical determinant of the survival and vulnerability of the prey. This portion of the model is a probabilistic matrix which mimics real weather records for the park.

The second portion of the model treats the dynamics of the prey species. Moose and Dall sheep are assumed to live in the territory of the pack all year; highly nomadic caribou move in and out. This translates into a set of dynamic equations for the moose and sheep which divides their life history into nine age classes. Caribou are treated as an external forcing factor, i.e. they influence wolf numbers but wolves do not influence theirs. Survival and reproduction of the prey species are reduced by crowding as well as by predation.

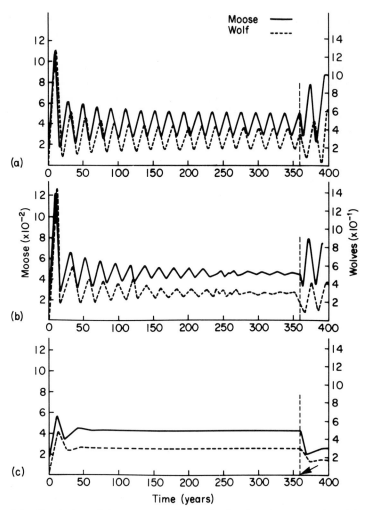

Figure 2.3 Simulation of the moose–wolf model reflecting the assumption either that (a) wolf reproduction was constant, or (b) wolf reproductive success varied with winter survival. Graph (c) reveals the stabilization that results from the presence of garbage dumps as an alternate food source. The vertical dashed line indicates the onset of hunting (from Crête *et al.*, 1981a).

The third submodel deals with the dynamics of the wolves themselves. Haber divided the pack into a core of dominant individuals and a fringe of younger subordinates. The core group is responsible for maintaining the pack's social structure and determines its hunting rate. Haber's model allows both for increased mortality from starvation when the pack is large and also for the formation of a smaller pack by fission.

The final submodel treats the interaction of the wolves and the several prey. The form of the mathematical description is a simple extension of a model I shall present in Chapter 8.

Simulation studies of the model pointed to several assumptions as being particularly important. Such things as the flexibility of the size of a pack's territory or the sensitivity of moose reproduction to population density proved to be influential determinants of population dynamics and worthy of further empirical research. A number of interesting predictions emerged. Wolf predation may have little influence upon ungulate populations until those populations are reduced by over-harvesting or by natural catastrophes. At such times reduction of wolf numbers may be necessary to prevent the system from crashing. Figure 2.4 shows one simulation in which an excessive moose harvest triggered a catastrophic decline first in moose, then in Dall sheep, and finally in wolves.

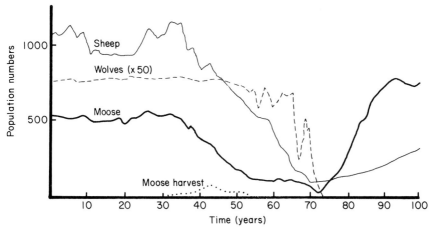

Figure 2.4   Model of Alaskan wolf–ungulate interactions simulated under circumstances in which an excessive harvest of moose triggered a catastrophic decline in both predator and prey (from Haber, 1977).

The African, Canadian and Alaskan studies all emerged from the desire to understand a particular ecosystem. They interdigitate nicely with accompanying empirical research and either have become, or are likely to become, the foundations of useful management models. Their construction is typical of a new and promising approach to applied problems in ecology. But while extolling such models, I must also note their limitations. To understand them, particularly the wolf–moose models, one must expect to spend a week or two in front of a computer terminal. And no such model is likely to be useful in circumstances other than those for which it was created, not even for the same species embedded in another ecosystem. Although all are predator–prey systems, none tell us much about what we should expect from predator–prey systems in general. For answers to broader questions, we must turn toward another class of models, those which I shall loosely classify as general theory.

General theory in ecology is the theory of a process reduced to its essence, stripped of the ornamentation of natural history. The subtleties that make each ecosystem unique should not be sought in either the construction or the predictions of such theory. Just as theoretical physics is not asked to predict precisely how far a tumbling boulder will roll before it stops, theoretical ecology should not be expected to produce accurate predictions of the function of real ecosystems.

Two important reasons exist for the field ecologist to study general theory. The first is practical; if a detailed model becomes necessary, its structure will almost certainly emerge from consideration of a general model. The second is more abstract but no less important. Perhaps more than other biologists, ecologists rely upon theory to focus and structure observations of their subject matter. To that end, general theory provides us with a minimum set of classes of population dynamics to plan for in the experimental design and look for in the analysis of data. Why is this true? Any field system is capable of being modelled, at least in principle, and that specific model, however large it may need to be, will contain the simple general model. Therefore, a larger model must be capable of exhibiting, as a minimum, the complexity of dynamical behaviors shown by the general model from which it emerged. From this line of reasoning it follows that field biologists who intend never to write an equation in their careers profit nonetheless from study of the basic theory. Development of a reasonable set of hypotheses to motivate an empirical research program requires it.

The remainder of this chapter comprises a simple disquisition on modern predator–prey theory. As I work through it, you will notice processes and concepts employed in the three wildlife models. In general terms the approach is this: the state of a system of n components reveals itself as a point in n-dimensional space, each axis of which represents the size of one component. These components or variables would usually number two – the predator's numbers and those of the prey – although this is not necessary; the variables could be aggregates or subdivisions of populations. The n-space is the set of combinations of all possible values of these components. In this scheme, the traditional population ecologist chooses to work in one-dimensional space, that single axis representing the number of prey organisms. Consider, for example, the time-series data plotted in Fig. 2.3(b). These could easily be represented as in Fig. 2.5, a graph of the number of wolves against the number of moose. The trend toward stabilization is revealed as a spiraling inward toward a single point. Of course, if one works with only one species then one has to find something against which to plot numbers, and time is the logical choice. The ecologist's goal is to understand the pattern of movement of the point representing the state of the system in n-space through time. A commonly-chosen tactic to approach this goal is to insert sets of specific values

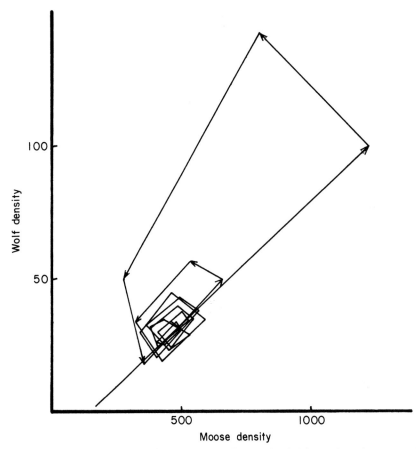

Figure 2.5   Wolf–moose simulation results from Fig. 2.3(b) replotted on a wolf density–moose density plane.

of the component variables, sets of points in *n*-space, into the model and associate each point with a direction and magnitude of change in each variable. The computer simulations of the three wildlife models employed this tactic. But point-by-point analysis is rarely satisfactory in the development of general theory; even if the possible number of states of the system are finite they almost always exceed the practical limits of this sort of analysis. So theoreticians have developed a few tricks to help with this task. The first one is to try to divide the *n*-space into regions (subsets of points) within which all the behaviors of the system are qualitatively similar. The task is to identify the boundaries of such regions and the qualitative behaviors characteristic of points within them. This seems mysterious only because of the jargon; the field population ecologist has precisely the same goal. He wants to know how a population will change from any given density, whether some densities predispose a population to extinction and others to outbreak.

The first steps in analyzing a model are routine and mathematically well understood. One solves for sizes of the component variables at which those variables do not change. If sets of values of the components exist which, when substituted into the model, yield no subsequent change in any of the system's components, then these we call equilibria. Equilibria in n-space provide valuable reference points to help partition that space into zones of similar dynamical behavior. If the system tends to move toward and remain at an equilibrium from within a region around it, then that equilibrium is considered stable. The zone of attraction may be small or it may comprise the whole of n-space. The equilibrium may not attract at all; it may repel the system but still be of interest because it helps subdivide n-space.

The mathematical concepts of equilibria and their stability have historically exerted a strong influence upon population ecology. The importance of population regulation as a research topic derives from the early fusion of this mathematical approach with the paradigm of the balance of nature, a concept which, for largely nonscientific reasons, is deeply embedded in the Western mind (Egerton, 1973; Baumer, 1977). The strength of this fusion generated a classical Hegelian dialectic. A group of ecologists arose who argued against the existence of equilibria in populations, generating the 'density-dependent versus density-independent' debates which occupied much of population ecology for a couple of decades (reviewed by Clark *et al.*, 1967).

Nature's balance or lack of it is not critical here. The necessary question is whether the qualitative features of Nature's operation can be modelled with sets of interacting equations of the most general and inclusive form. If not, then you may close the book and proceed no further; the field of ecology is not a scientific discipline.[2] If so, then the procedures of solving for equilibria and determining their stability are useful mathematical first steps. I emphasize that they are only the first steps toward understanding the dynamics of an ecological system over the whole of n-space.

To bring this down from the clouds a little, I offer the following model, a set of two ordinary differential equations. The population size of the predator is given by P; that of the prey is given by H (for hosts, herbivores, or whatever). As customary, the model is posed as descriptions of the rates of change of the two populations.[3]

$$\frac{dH}{dt} = Hg(H) - f(P, H) \qquad (2.7)$$

$$\frac{dP}{dt} = k(P, f(P, H)) \qquad (2.8)$$

Equations (2.7) and (2.8) actually comprise a class of models, since the forms of the functions $g(H)$, $f(P, H)$, and $k(P, f(P, H))$ are not specified. To proceed with this model I must put a few restrictions on these functions.

The function $g(H)$ in Equation (2.7) is the growth rate per individual of the prey population in the absence of predation. This will be assumed to be maximum at some low density of prey and reach zero as the prey approach the limits of their resource base. Note that in some circumstances, such as the first two wildlife models, $g(H)$ could be simplified even further to a constant rate of increase per prey individual, independent of density. The negative term in Equation (2.7) represents the rate of loss of prey organisms to predation. For a fixed value of P the function $f(H, P)$ describes the dependence of the rate of killing per predator on the density of prey. Normally $f(H, P)$ increases to some maximum value set by the available time, the maximum rate of digestion, or whatever limits the predator's killing rate. If more predators exist, they kill more prey. The growth rate of the prey population is the difference between the first term, which will usually be positive, and the second term, which is negative or zero.

The remaining function, $k(P, f(P, H))$, is the growth rate of the predator population. Its arguments appear as P and, separately, $f(P, H)$ to reflect that growth in the population of predators depends upon both the density of predators and the rate of food supply, $f(P, H)$. The rate of population growth will tend to increase as the rate of killing increases up to a limit determined by the tolerance of predators to crowding.

Although simple, this model is sufficiently realistic to justify serious scrutiny. The analysis of such models as this follows certain customary procedures, the goal of which is to be able to predict the subsequent trajectory of the system starting from any point on the H–P plane. The first step in realizing this goal is to find conditions for cessation of change in the prey and predator populations. The prey population, governed by Equation (2.7), will not change when $dH/dt = 0$, or

$$\hat{H}g(\hat{H}) - f(\hat{P}, \hat{H}) = 0. \qquad (2.9)$$

Densities which satisfy Equation (2.9) will be designated with a circumflex. Two such sets of values of $\hat{H}$ and $\hat{P}$ are obvious: if $\hat{P}$ equals zero, then $f(\hat{P}, \hat{H})$ must also be zero. It follows that Equation (2.9) holds for both $\hat{H} = 0$ and $g(\hat{H}) = 0$. These are not terribly interesting conditions, since in the first neither predator nor prey exist and in the second the predator is absent, but they prove important in structuring the analysis of the model. Other values of $\hat{P}$ and $\hat{H}$ exist which yield no growth in the population of prey. The set of such values could describe the curve in Fig. 2.6. The curve divides the plane into a region below, in which the sign of the derivative in Equation (2.7) is positive and the prey increase in numbers, and a region above the curve, in which the derivative is negative. The curve, which is called the prey isocline, is said to divide the plane into 'phases' and thereby lends to it its sobriquet, the 'phase plane'. The important feature of the curve is the hump. Its existence has been justified on both theoretical and empirical grounds and will not concern us here,[4] except to note its interpretation. At the density of prey given by the peak of the isocline, the population of predators must be abundant to prevent growth of

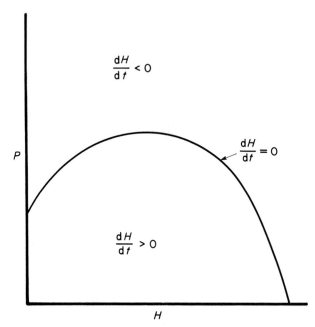

Figure 2.6 The curve describes the set of prey (H) and predator (P) densities which result in the absence of growth in the prey population.

the prey population. Smaller or larger densities of prey require smaller populations of predators to prevent growth. On the right side of the hump the prey crowd themselves sufficiently to curtail growth; on the left side they are too few to face the pressure of their predators.

Inasmuch as Fig. 2.6 was drawn from consideration only of the prey Equation (2.7), it reveals nothing of the conditions for cessation of population growth in the predator. When the derivative in Equation (2.8) is zero, then $k(\hat{P}, f(\hat{P}, \hat{H}))=0$. A reasonable form for the function k is the line in Fig. 2.7. At low densities of predators, their population growth or decline depends only upon the abundance of prey; at high densities, growth depends only upon their own numbers. If both the prey and predator isoclines are drawn on the same graph (Fig. 2.8), their crossing yields a set of population densities at which both species might coexist at equilibrium. With the finding of these equilibrium densities, the first step in the analysis of the model is complete. As drawn in Fig. 2.8, three equilibria exist: when both are absent ($\hat{H}=H_1=0$, $\hat{P}=0$), when the prey population is at saturation and the predator is absent ($\hat{H}=H_3$, $\hat{P}=0$), and when the populations are at the densities indicated by the intersecting isoclines ($\hat{H}=H_2$, $\hat{P}=P_1$).

The stability of these equilibria is revealed by the second step in the analysis. Using the mathematical process of linearizing around an equilibrium,[5] one can determine if very small departures from the equilibrium point vanish or

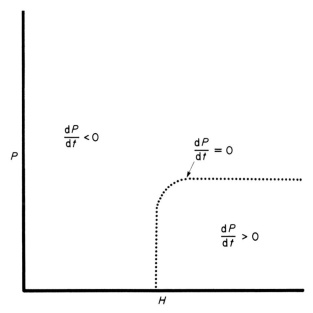

Figure 2.7 The dotted line gives the values of prey and predator densities yielding no growth in the predator population.

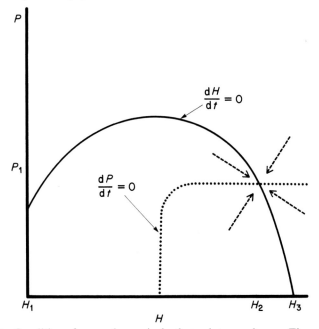

Figure 2.8 Conditions for no change in both predator and prey. Three equilibria exist: H_1 has neither prey nor predator; H_2 has both; H_3 has only prey. The dashed arrows reveal the trajectories of populations near the single stable equilibrium, $\hat{H}_2$.

grow larger; whether, in other words, the equilibrium attracts the system or repels it. The definition of stability as attraction by an equilibrium is the common use of the term in ecology as well as mathematics. Of the three equilibria in Fig. 2.8 only one, (H_2, P_1), is stable. When near to it, the system will proceed directly to equilibrium. The zero equilibrium is stable to perturbation in the direction of the P-axis but unstable to any perturbation involving addition of prey. The largest equilibrium $(H_3, 0)$ is stable along the H-axis but unstable to any perturbation involving addition of predators.

Suppose that a different predator is involved – one that is tolerant of crowding but requires more food for population increase. We might get a phase-plane diagram as in Fig. 2.9. Here, as before, the equilibria are three in

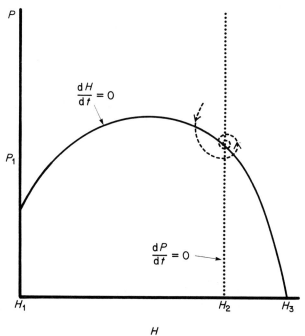

Figure 2.9 Oscillatory convergence to a stable equilibrium when the population of predators is not limited by intraspecific mechanisms.

number, and only (H_2, P_1) is stable. But it is stable in quite a different way. The system will return to it if perturbed, certainly, but it will return by both populations overshooting and undershooting the equilibrium, oscillating with decreasing amplitude.

An important point that emerges in comparing Figs 2.8 and 2.9 is that the absolute size of an equilibrium population in the presence of predation may not reveal much about the expected pattern of fluctuation about that equilibrium. Note also that the prey densities at the two equilibria, H_2 and

H_3, are not substantially different. In the face of counting errors and environmental fluctuation they might not be distinguishable.

An understanding of the local stability of the point equilibria does not, by any means, complete the analysis of Equations (2.7) and (2.8). The next step is to try to see how much this stability analysis tells us about the model's behavior in the remainder of the phase plane. A few tricks help here, but most biologists, even those with a theoretical bent, resort to computer simulation for this phase of the analysis. For some models mathematical procedures exist that enable us to say something about the whole phase plane; fortunately the graphs in Figs 2.8 and 2.9 describe such a model. By application of these procedures we know that the locally stable equilibrium is also globally stable; no matter what the starting populations are, the system should end at (H_2, P_1) (Gatto and Rinaldi, 1977; Hastings, 1978; Hsu, 1978).

What would happen if we tried a third predator with different properties, a more efficient predator limited only by food supply? The phase-plane might look, in this case, like Fig. 2.10. Analysis now reveals that all three equilibria are unstable. Small perturbations away from (H_2, P_1) must result in the system going somewhere, of course. And what happens is this: these perturbations increase in amplitude in an oscillatory fashion, but after winding out from (H_2, P_1) the amplitude of the oscillations ceases to grow and the system settles into a steady cycle. This 'limit cycle' is stable in its own fashion. The system will

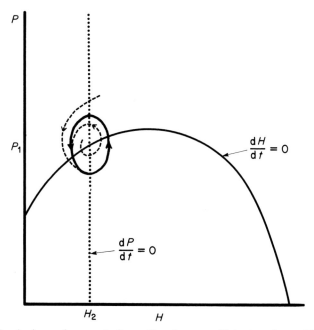

Figure 2.10   A phase-plane portrait resulting from an efficient predator. All equilibria are unstable. Populations converge to a stable limit cycle.

circle into it from the outside as well as the inside.[6] At first glance this seems a peculiar thing for a population to do, but we shall see experimental evidence for this behavior in Chapter 10.

Limit cycles arise often in predator–prey models. One such example is the continuing oscillation in Fig. 2.3(a). The amplitude of the limit cycle is a function of how far to the left of the hump H_2 occurs. Close to the hump, the oscillation is small (Freedman, 1980). As H_2 gets smaller the cycle increases in amplitude until the predator and prey populations fluctuate so wildly that by chance one or the other goes extinct.

In summary, this model (one of the simplest practical models of predation) exhibits a variety of dynamical behaviors. It displays stable equilibria in which populations proceed smoothly toward constant sizes; it displays stable equilibria in which populations oscillate before settling down to constancy; it exhibits equilibria which are not stable at all but which are surrounded by stable limit cycles of a fixed period and amplitude.[7] The model suggests in addition that the influence of a predator upon the equilibrium abundance of its prey is not an unambiguous guide to its influence upon the stability properties of that equilibrium, or, to state the converse, how a predator regulates its prey is not a reliable guide to the equilibrium density of that prey. Such a relationship between equilibrium and stability may exist for a particular model; if so, it is specific to that model and is not a general feature of predator–prey systems.

If the claim I make is true, that the simple general theory can be discovered in more complicated and specialized models, then it must follow that the spectrum of mathematical behaviors I have just described comprise only a subset of the behaviors of realistic models of predator–prey interactions. These kinds of dynamics must, by extension, be considered the minimum possible set in any field situation until some of them are demonstrated not to occur by research.

The temptation will be strong to relegate the procedures I have outlined to a sort of tool box of mathematical tricks which can be ignored until one needs the appropriate gadget to solve a specific problem. This would be a great misapprehension. The primary value of general theory derives from the logical insight it provides into the workings of the ecological world. To illustrate this point I devote the next chapter to some of the conceptual problems that have emerged from verbal predation theory, problems both historical and current.

NOTES

1. This deficiency is in the process of being redressed. See Starfield *et al.* (1981) and the references therein.
2. In the words of a recent Nobel laureate:

    The general paradigm is: given a blueprint, to find the corresponding recipe. Much of the activity of science is an application of that paradigm: given the description of some natural phenomenon, to find the differential equations for processes that will produce the phenomena.

Simon (1969, p. 112)

3. Entomologists have generally favored difference equation models for highly-specific insect parasitoids, in which the host and parasitoid breed synchronously. See Hassell (1978) for a review of this work. Whether one uses difference or differential equations for purposes of general theory is not nearly so important as using a model which is sufficiently robust to accept the reintroduction of some of the more important details of the natural history of the species involved. The classical predator–prey equations of Lotka (1925) and Volterra (1928) provide a well-known example of an insufficiently robust model. Their prediction of continuing neutral oscillations proved to be pathologically sensitive to the relaxation of some of the model's most restrictive assumptions.

4. See Chapters 5 and 8; also Rosenzweig (1969). The phase-plane approach to a general model was pioneered by Rosenzweig and MacArthur (1963).

5. The mathematical procedures involved will not be covered here. Maynard Smith (1968) presents a simple portrait of this process; Vandermeer (1981) and Pimm (1982) go into it in more detail.

6. This is known from the Poincare–Bendixson Theorem. Limit cycles in predator–prey systems were originally pointed out by Kolmogorov (1936, cited by Roughgarden, 1979), although most ecologists were unaware of them until articles by Canale (1970), May (1972) and Gilpin (1972).

7. These dynamical behaviors by no means exhaust the range of possibilities. Theoretical ecologists have, of late, been caught up in the world of strange attractors and chaotic or pseudorandom dynamics. I have chosen not to treat these mathematically embryonic topics here. If the assertion that 'the degree to which real ecosystem behavior is chaotic is possibly the most fundamental question facing community ecology' proves true (Gilpin, 1979), then this and a number of other books in ecology will need to be rewritten.

3 Clearing the decks

I have an old belief that a good observer really
means a good theorist.

Darwin[1]

The fundamental question to which ecologists have addressed themselves
since the origin of interest in predation is what functional role predators play
in the community. Are predators important to prey populations or are they
not? Since these are somewhat vague concerns, the earlier workers sought to
pin the topic down by asking more specific questions, the most common of
which was: do predators control their prey?

Before examining some of the answers put forth, we might be well advised to
figure out what the question asks – to discover, in other words, what is meant
by control. The term control turns out to be one of ecology's great buzzwords,
the sort of term that Hardin dubbed a 'panchreston', a word which can mean
anything and, by extension, usually means nothing (Hardin, 1959). To
illustrate this, G. C. Varley (1975) created an abstract of a fictitious paper in
which *control* is used six times with six different meanings:

Control of the bean bug in Ruritania
Abstract. Damage by the bean bug is variable in Ruritania and weather is
mainly responsible for its control[a], but the level of control[b] on the coast is
unacceptable because up to 50% of the crop may be unsaleable.
Experiments using DDT to control[c] the pest proved ineffective, even
though control[d] was repeated monthly and over 90% control[e] was
achieved. Pest resurgence was rapid and the bug population was soon
above that in the control[f] area where no insecticide was used.

All of Varley's six uses of the term have appeared in the literature. They have
the following precise meanings: (a) a controlling factor is that which accounts
for the greatest mortality in the population; (b) control implies regulation of
the population about some equilibrium; (c) a population under control is one
kept below an economic threshold; (d) a control effort is equivalent to an
application of a pesticide; (e) control is identical to mortality; (f) controls are
comparisons to treatments in scientific experiments. This set of six definitions
is unfortunately not exhaustive of the term's use.[2]

To put such an ill-defined word into the statement of a theory is to make that theory untestable. Recognizing this, various groups of predation workers have tried to define control more precisely. Inevitably, working as they have in isolation from one another, these groups have opted for different precise definitions. Entomologists in the field of integrated pest management usually use the third definition: the keeping of a pest population below an economic threshold (Huffaker *et al.*, 1971). Wildlife biologists typically employ the second, e.g. a deer herd is controlled if it is stabilized. To the researcher reading in both fields, the confusion remains.

On avenue of escape is to avoid the use of the word entirely; one merely addresses a different specific question. A popular alternative is to ask if predation is the limiting factor for prey populations. Again we might profit from dissecting the question and address the definition of a limiting factor. This term arose from one of the great conceptual advances of the early days of ecology, the 'Law of the Minimum'. J. Liebig claimed that no matter how many nutrients plants use, their growth ceases with the exhaustion of the supply of just one. That nutrient in shortest supply was designated the limiting factor. The influence of this idea is hard to over-estimate. It greatly simplified the job of the ecologist; rather than investigate the entire environment of an organism, one had only to identify that single limiting factor. Equally important, the experimental approach was straightforward: one manipulated the factors influencing a population one by one and identified as the limiting factor that which caused a response.

The elegantly simple Law of the Minimum became dogma for any question dealing with the regulation of a population. And as it became dogma its advocates fell into a trap characteristic of most conceptual advances:

> When one becomes aware of and accepts a set of new ideas, there is the tendency to overgeneralize their significance, i.e., to perceive the world in terms of the new ideas, to restructure perceptions, to reinterpret past experience, and to see an altered future. It is a mixed blessing. On the one hand, there is the possibility that the ideas have validity, if only to the extent that other people feel compelled to take them seriously; on the other hand, they may be so idiosyncratic (however compelling to you) that you are unaware you are applying them far beyond their merit and appropriate boundaries.
> (*Sarason, 1977*)

The appropriate boundaries for Liebig's Law are not easy to specify but probably do not extend much beyond plants whose growth is limited by the availability of inorganic nutrients. At least we know that animals which eat other living organisms can violate the basic prediction of the Law, that manipulation of one factor alone will produce a response in the population (Wagner, 1969).

The truth of this is revealed by examining the phase-plane portrait of a simple model of the sort described in the previous chapter. In Fig. 3.1, prey and

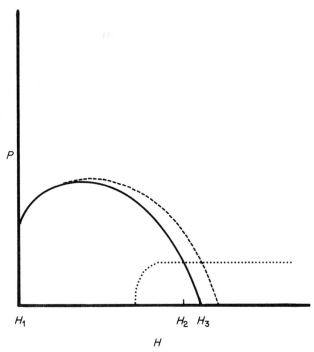

Figure 3.1   Phase-plane portrait of a circumstance in which either the elimination of predators or the enhancement of the food supply of the prey would shift the equilibrium from H_2 to H_3.

predator isoclines are drawn which reflect the effects of self-limitation on the part of both populations. If one were to enhance experimentally the supply of food to the prey population, one could shift the prey isocline from that represented by the solid line to the dotted line. The new equilibrium density of prey would clearly be larger than the original. But had the prey's resources been left alone and the predator population experimentally removed, the equilibrium density of prey would also have increased, in this case to the density represented by the intersection of the solid prey isocline with the *H*-axis. Since the two responses would be indistinguishable, the experimenter would have difficulty inferring which is the limiting factor; both generate a response when one is manipulated and the other is held constant as a control. Obviously ecologists did not notice the synergism between food supply and predation when applying Liebig to predation. But the relationship makes meaningless the use of limiting factor in this context.

Construction of one counter-example to the Law of the Minimum is relatively simple. Substantially less simple is the separation of general classes of ecological processes into those subject to the Law and those not. The simple relationship between resources and population abundance implied by the limiting-factor approach apparently holds only for ecological systems that are dimensionally simple. What makes for dimensional simplicity? Armstrong and McGehee (1980) have made some progress toward answering this question by noting that the renewal of some resources can be adequately described by algebraic equations (see also Schaffer, 1981). For example, if a consumer population of size N grows in direct proportion to its consumption of some resource available in quantity R, then we might describe the dynamics of N by

$$\frac{dN}{dt} = aNR \tag{3.1}$$

$$R = R_{max} - cN \tag{3.2}$$

where a and c are appropriate conversion constants. Armstrong and McGehee call R a conservative resource because its total size, R_{max}, is always constant. The proportions available, R, or tied up by consumer individuals, cN, can vary certainly, but they must always sum to R_{max}. If one substitutes the second equation in the first, one gets

$$\frac{dN}{dt} = aR_{max}N\left[1 - \frac{N}{R_{max}/c}\right] \tag{3.3}$$

which is the logistic equation for $r = aR_{max}$ and $K = R_{max}/c$. Their point is that this, the most commonly used model for population growth, derives from a special portrait of how resources are renewed, a portrait which is consistent with Liebig's Law.

One is hard pressed to generate a reasonable argument for the use of Equations (3.1) and (3.2) when the resource is a biological species. Biological populations are not conservative; their numbers are not constrained to remain at a fixed density. To build a reasonable model of a population exploiting a biological resource, one is forced at a minimum to the sort of model described in the previous chapter, in which the prey population is given its own differential equation. The inevitable consequence is greater dynamical complexity than the limiting-factor approach can accept. In summary, Liebig's Law of the Minimum applies *only* to one-dimensional processes of the type adequately described by Equations (3.1) and (3.2), and predation is, at a minimum, two dimensional.

I have used these two problems to illuminate common, perhaps inevitable, weaknesses in verbal theory. On one hand terms can be ill defined; on the other the logical basis for an argument is sometimes so cryptic as to be inaccessible. The usual consequences are confusion and incorrect reasoning. To further bolster this point, I shall devote a few paragraphs to the ideas of the greatest of the non-mathematical predation theorists, Paul Errington.

Errington came into wildlife biology at a time when its theory of predation was simple: if predators eat prey, the removal of predators means more prey.

His field work on bobwhite quail and muskrats convinced him this was untrue. In a number of books and papers, Errington argued that the sizes of prey populations are determined in many cases by behaviors which are intrinsic to those populations, such as territoriality. His observations of differential vulnerability of social dominants and social subordinates suggested that predators took prey which would have died anyway from disease or starvation.[3] This 'doomed surplus' was functionally unimportant to the prey population. The argument was a major conceptual advance, recognizing as it did the interactive and compensatory role of predatory and other kinds of mortality. Any sophisticated theory, verbal or mathematical, must include this much of Errington's ideas. And by reducing predators to the ecological equivalent of garbage collectors, Errington undoubtedly served to forestall the conscious eradication of a number of carnivorous birds and mammals from North America. Perhaps the greatest testimony to his impact is that several decades after his work many, if not most, wildlife biologists hold his ideas to be inviolate.

Criticisms surfaced every now and then, of course. Craighead and Craighead (1956) asked why, if predators removed surplus animals, they should not be considered to be regulating the prey? How prey die is important, they asserted. This argument was ignored in the face of Errington's focus upon the extent to which predation was critical to the process of population regulation. Would muskrats reproduce and survive to alarming densities if all mink were removed from a marsh? Errington's answer was no, of course not; therefore, predation was not necessary to population regulation.[4]

Errington was without question a devotee of the limiting-factor approach and, in common with those wielding Occam's razor, looked for the single factor limiting population growth.[5] Yet he restricted his discussions of limiting-factors solely to the question of stability or regulation. From the previous chapter we know that the influence of predators, be it strong or weak, can be evidenced in both the stability of a prey population and its equilibrium abundance. We know, in addition, that neither influence need relate quantitatively to the other.

By restricting his focus solely to population regulation, Errington made some curious arguments. He defined the 'doomed surplus' on a number of occasions as those prey which exceed a threshold of security. The threshold was defined by the availability of suitable habitat, and suitability, in turn, included protection from predators. Errington would assert at one point that

> we may see that a great deal of predation is without depressive influence. In the sense that victims of one agency miss becoming victims of another, many types of loss – including loss from predation – are at least partly intercompensatory in net population effect.

And then a paragraph later he would follow with

> Unanswered questions also remain as to what proportions of the habitats

that are marginal for various prey species might accommodate greater proportions were it not for interspecific predation.
(*Errington, 1946*)

Apparently predators are allowed to influence the equilibrium abundance via defining suitability of a habitat type and yet not qualify as a limiting factor. One might accept that predators do not play a critical role in the regulation of prey populations, but if the doomed surplus did not support local predators, would it not follow that the prey could expand into marginal habitats? The mode of death in this circumstance is quite important.[6]

Nobody should fault Errington's major contribution. He eliminated a great deal of simplistic reasoning by pointing out the logical fallacy of deducing a major population impact from a frequent and dramatic behavioral event. But his attempts to articulate an alternative theory of predation bogged wildlife biologists down into a morass of tortuous verbal logic, a morass from which they are just now emerging and which would have been quickly and cleanly bridged by the mathematical theory.

NOTES

1. From an 1860 letter to Bates, cited in Ball (1975).
2. See, for example, Keith (1974) in which control is defined as 'the maintenance of a population in being; such maintenance may involve both density-dependent and density-independent processes'.
3. See Errington (1946, 1956, 1963a, 1967). His data on muskrat populations appears in Errington (1963b).
4. Bulmer (1974) argued from consideration of Canadian fur records that mink do depress muskrat numbers.
5. 'My own studies of such patterns have dealt with what are commonly thought of as limiting factors in mammal and bird populations.'
(*Errington, 1956*)
6. This somewhat cryptic feature of Errington's reasoning went unnoticed until 1970. See Huffaker (1970) and Watson (1970).

4 Field studies

Probably the commonest death for many
animals is to be eaten by something else.

Elton (1927)

All styles of ecological research share the goal of making predictions about the operation of real ecosystems, but how this is best done is not always clear. I thought it worthwhile, therefore, to begin discussion of the empirical literature by examining a number of case histories of respected field studies of predation. By exposure to their design and interpretation, you may be led inductively to consideration of some of the general issues that impinge upon the conduct of field research on this topic. The chapter has the secondary, but nonetheless important, purpose of laying to rest the antiquated belief that predators are an unimportant influence upon natural populations of prey.

The North American ecologist is inevitably drawn first to what has become a classic in the predator–prey literature: the wolves and moose of Isle Royale (Fig. 4.1). Isle Royale lies in Lake Superior about 30 km off the northwestern shore. It and its surrounding small islands form the Isle Royale National Park, with an area of about 540 km². Moose arrived on the island in the early 1900s and subsequently increased to high densities. The early records of their numbers are not accurate, but the impression is one of great fluctuation and instability with numbers rising in excess of 2000. In the late 1940s wolves invaded the island and began preying upon moose. Shortly thereafter students of D. Allen at Purdue University began a series of descriptive investigations of this interaction, a series which has made the wolves and moose of Isle Royale the best understood ungulate–large carnivore system in the world.

The first and most well-known of these studies was done by David Mech (1966, 1970), who estimated densities of wolves and moose, reproductive rates of moose and killing rates of wolves. From these data, Mech concluded that wolves were regulating moose on Isle Royale, that they destroyed a sufficient number of animals each year to maintain the population at equilibrium. Mech's study was a critical first step to understanding this system, but, based as it was upon a purely descriptive study over a short interval, its conclusions did not hold up in the face of later evidence. Jordan *et al.* (1967), working several years later with improved census techniques, found that the numbers of moose were 50% larger than the earlier estimate, around 900 rather than

Figure 4.1 Wolf packs on Isle Royale (a) on the move, (b) in pursuit of prey, (c) harassing a moose at bay (photographs by R. O. Peterson).

approximately 600 (Fig. 4.2). Either the original estimate was too low, in which case Mech's estimate of the production of moose should have exceeded the rate of killing, or the original estimate was good and the wolves had failed to stop population growth. The latter is more likely (Mech and Jordan, personal communication). Subsequent studies by Wolfe and by Peterson have documented a surprisingly complex relationship between this predator and its prey.[1] Moose numbers peaked in 1967 at about 1500, only to collapse shortly thereafter to low densities throughout the 1970s. Except for a period of instability in 1967 and 1968, wolves held together in a single pack of about 22 animals for over a decade. A second pack appeared in 1972, and a third pack emerged in the winter of 1974–1975. Following this split of the Big Pack, the numbers of wolves quickly doubled to about 44. The predation rate per pack remained about the same during this interval, resulting in a trebling of the rate of killing. The wolf population peaked at an astonishing density of about 50 in five packs in the winter of 1980 and then sharply declined in the winters of 1981 and 1982 to a low of 14 animals. At present both populations appear to be on the rise again. The abrupt decline in numbers of moose in the early 1970s resulted from a series of severe winters in 1969, 1971 and 1972, which resulted

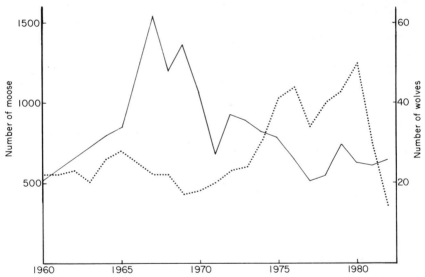

Figure 4.2 Population fluctuations of moose (solid line) and wolves (dashed line) on Isle Royale. These numbers are quite accurate for wolves. The numbers of moose were arrived at by aerial census (Jordan and Wolfe, 1980) or by averaging three independent estimates of abundance, of which one was an aerial census (Peterson and Page, in press; Peterson, personal communication). The estimates of moose abundance are somewhat higher than those given by Peterson and Page because I weighted the aerial counts by the 16% figure given by Jordan and Wolfe as the approximate proportion of the population missed by this census tactic.

in the production of a weak age cohort. Mech had discovered earlier that wolf predation was selective for calves or old, infirm adults. But the bad winters of the early 1970s resulted in an unusually high percentage of predation upon young adult moose, those in the 1–3 year age class.

Obviously the wolf–moose relationship is not as predictable as originally anticipated. Mech's simple tactic of discovering regulation by balancing the rates of prey production and predator killing looks a bit naive, given the benefits of hindsight. The moose population appeared to change only partially in response to wolf predation; a major factor throughout the 1970s has been the abundance and availability of winter browse. The predatory impact of wolves is in turn linked not only to moose density but also to pack structure, the social dynamics of which may have led both to the long interval in the 1960s during which wolves did not respond to increases in moose density, and to the explosive increase in the early 1970s after the pack split. Whether the system will continue this oscillation is not clear. One must conclude that the best understood carnivore–ungulate system in the world is only now becoming well enough researched, after twenty-five years of intensive observation, to allow a few tentative predictions to be made of its future course.

Most field ecologists have substantially less in the way of time and resources than the group working on Isle Royale and have, consequently, turned to a variety of experimental approaches for assessing predation's impact. The two most common techniques have been the addition and removal of predators – manipulations which are certainly instructive but are not the only or even, in some cases, the best experimental tactics. Nonetheless such studies comprise the bulk of the evidence we have for the functional role of predators and merit detailed consideration.

The most spectacular examples of the impact of predators come from the literature on the biological 'control' of insect pests. Of the many attempts to minimize pest damage by use of natural enemies, I have selected two for presentation here: the olive scale (Heteroptera: Diaspididae) and the cottony-cushion scale (Heteroptera: Margarodidae).[2] The olive scale, a pest of many trees and shrubs, first appeared in California in 1931. It quickly became a serious pest on olives, deforming the fruit and killing limbs. Starting in the late 1930s, entomologists introduced a variety of predatory insects to reduce infestations of this pest. They were finally successful in the 1950s with an aphelinid wasp, *Aphytis maculicornis*. It spread rapidly and caused a sharp decline in scale infestations. Unfortunately *A. maculicornis* proved vulnerable to hot, dry summers and did not reliably maintain the olive scale at low levels. The introduction of a second aphelinid wasp, *Coccophagoides utilis*, around 1960 achieved the desired economic control. In concert, these two wasps have maintained the olive scale stably at an economically low equilibrium.

The cottony-cushion scale, an older pest in California, first appeared in 1868 from its native Australia. This insect attacks a variety of trees and shrubs; in

California it appeared on citrus at the beginning of that state's profitable citrus industry. It had a devastating effect, killing large branches, even whole trees. In the late 1880s several potential enemies were imported, among which was the vedalia, a coccinellid beetle. The vedalia was remarkably successful over the course of only a few years, reducing the cottony-cushion scale to economically unimportant levels. An agromyzid fly, *Cryptochetum iceryae*, has also become established and tends to be the more common enemy in cooler coastal areas.

To object that neither of these two examples provides a tidy, well-designed experiment to ponder would be true but quibbling over details. The patchy nature of the action of these predators provided an abundance of natural experimental controls, so that we can be confident that the declines in the two species of scales can be attributed to their predators. Even now outbreaks of the cottony-cushion scale follow hard winters or pesticide applications which destroy the predators. Such outbreaks are easily stopped by the introduction of as few as five pairs of vedalia beetles.

Predator-removal studies fall into two classes according to their length and geographic extent. Most are short, by comparison to the generation times of the animals, and cover only a portion of the range of a population. Relatively few last several generations and influence the complete range. It should be said in their defense that short predator-removal projects almost always reflect efforts by applied biologists to explore one particular management tactic: most were never intended to reveal anything fundamental about the predator–prey relationship. I shall illustrate the approach by discussing five studies which show an effect of predation.[3]

The first is a study by Blankinship (1966) of predation on white-winged doves. White-winged doves nest in dense breeding colonies in citrus groves and remnants of woodland tracts in the lower Rio Grande Valley of Texas where their nests are exposed to predation by the great-tailed grackle. During the spring and summer of the years 1964 and 1965, grackles on a 15.5 acre nesting colony of doves were shot. A total of 2824 were removed in the first year and 1398 in the second. The number of nesting doves, in response, increased to 449 pairs in 1964 and 529 pairs in 1965, as compared to the previous high of 256 in 1956. The colony fledged 377 young in 1964 and 358 in 1965, increases of about 200% over the best year before removal of grackles. Other colonies of doves in the same years, 1964 and 1965, showed only about 5% nesting success. The simultaneous controls were not quite comparable, so Blankinship's major comparison was to the same colony during the years before experimental treatment.

In the second study Chesness *et al.* (1968) investigated the influence of predation on the production of pheasant chicks in southern Minnesota during the years 1960–1964. They removed small carnivores from an experimental area of 2560 acres and maintained a comparable control area of 4080 acres. Pheasants nested in similar densities on both areas. On the trapped area, hatching success improved during the course of the study, reaching a high of

36%. On the control area it remained at 16%. The clutch size was also larger in the experimental area, a result which they attributed to enhanced survival of the larger first clutch. The net result was a rate of production of chicks in 1961 and 1962 on the trapped area about twice that of the control area, a difference that might have been even larger had domestic dogs and cats been removed.

Duebbert and Kantrud (1974) examined the nesting success of upland ducks during the summer of 1971 in north-central South Dakota. The predators of interest in the area were red foxes, raccoons, striped skunks, and badgers. The study sites were either grass–legume fields which had been idle for five or six years or active crop and pasture land. Experimental control and predator reduction areas were established in each type of habitat. Table 4.1 displays the results. Active agricultural land produced few ducklings, and no

Table 4.1 Duck nesting success in north-central South Dakota, 1971. Data are the number of nests km^{-2}, the percentage of nests in which eggs hatched, and the eventual number of ducklings produced per hectare (from Duebbert and Kantrud, 1974).

Treatment	Habitat type	
	Idle grass–legume fields	Active cropland and pasture
Control	84	14
	68%	51%
	4.8	0.5
Predator reduction	299	12
	92%	85%
	22	0.8

significant effects of reducing predators could be seen in this habitat type. Idle land proved to be substantially more productive, and here the effects of the experimental treatment were substantial. Both the nesting densities and the percentage of nests in which eggs hatched were greater in those areas from which predators had been removed.

The longest of this set of experiments is Potts' (1980) ten-year study of the nesting success of partridge in Sussex. Table 4.2 displays brood production rates as a function both of the density of nesting pairs and the intensity of predator reduction. The major nest predators, fox, stoats, cats and carrion crows, were removed occasionally throughout the study area but only intensively on the 13.1 km^2 North Farm Shoot. A substantially larger fraction of the nests survived where predator management was intensive. Partridge appear to exhibit density-dependence in brood production in the presence of predators but not in their absence.

Table 4.2 Variation of partridge nesting success as a function of both nest density and the intensity of predator removal in a Sussex study area from 1968–1977. Predator removal efforts are measured in units of gamekeepers per km^2 (after Potts, 1980).

	Predator removal	
Nests/km^2	0.25	0.07
1–5		0.67 ± 0.10
6–10	0.7 ± 0.07	0.51 ± 0.02
11–15	0.67 ± 0.05	0.43 ± 0.03
16–20	0.55 ± 0.04	0.43 ± 0.05
21–25	0.54 ± 0.04	0.29 ± 0.02
26–30	0.61 ± 0.04	
31–35	0.71 ± 0.22	
Sample size	59	105

The most convincing of this class of experiments is the work of Balser *et al.* (1968) on nest predation of ducks in the Agassiz National Wildlife Refuge of northwestern Minnesota from 1959 to 1964. They divided the study area into two portions, one experimental and one control. From the experimental half they removed all mammalian nest predators, a total of 1342 raccoons, skunks and foxes over the course of the study. The fraction of successful nests was twice as great in the experimental area, and the density of ducklings was 60% greater (not 100% greater because of re-nesting in the control area). Interestingly enough, they reversed the experimental and control areas half-way through the study and observed a reversal in nesting success. A number of artificial nests were placed in both areas; predators destroyed 66% of these in the control area and only 19% in the experimental area.

For various reasons, none of these five studies produced a detectable effect upon the adult population. Because the geographic extent of the manipulations was limited, the effects of the experiments were locally obliterated each season. One can conclude with some confidence that predators on the nesting grounds take a substantial number of eggs and nestlings. But such information is only a partial answer to the question of whether removal of all such predators would increase the population of adult birds. The remainder of the answer lies in identification of additional sources of mortality at other ages and whether any of these kinds of mortality are intercompensatory. Without question, predictions as to the course of entire populations come more easily with relatively sedentary and easily manipulated organisms with a short generation time.

Huffaker and Kennett (1956) published some intriguing data on an acarine predator–prey system in California strawberry fields. The cyclamen mite, a pest of strawberries, is secretive and usually avoids substantial mortality from the pesticide parathion. By comparison the predatory mite, *Typhlodromus reticulatus*, suffers substantial mortality with each application of parathion. Figure 4.3 reveals the effect of removing *Typhlodromus* upon the dynamics of the cyclamen mite in three replicate plots. That influence is straightforward: the cyclamen mite explodes in numbers shortly after each application of

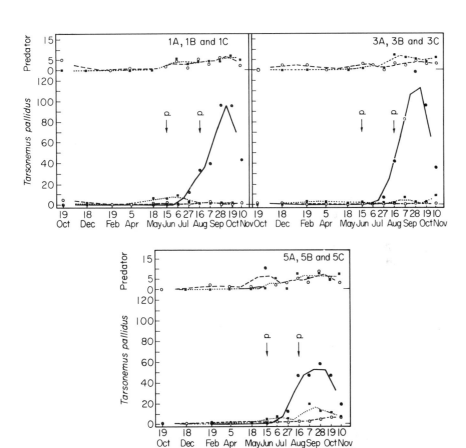

Figure 4.3 Three replicate experiments on first-year strawberry plants, 1953–1954. Each graph exhibits data from a control and treatments consisting of addition and removal of predators. Numbers of prey are per strawberry leaflet. Densities of predators are displayed as the number of leaflets in a sample of thirty containing at least one *Typhlodromus*. Each 'P' denotes a parathion application in the predator-removal treatments (from Huffaker and Kennett, 1956). ○ – – ○ Predator or prey in predator-stocked plots; ■ - - - -■ Predator or prey in unaltered plots; ●——● Prey in predator-free plots.

parathion. By virtue of its lack of numerical fluctuation we can infer that the population of cyclamen mites is stable in the presence of *Typhlodromus*.

Sedentary marine invertebrates comprise one of the groups of animals most amenable to experimental manipulation, and a number of quite good studies have been conducted on predation using intertidal and shallow subtidal species. The bulk of these are oriented to the structure of communities and do not relate directly to the subject of the book (see Chapter 12). I review here only two, the first by Connell (1970) and the second by Virnstein (1977).

Connell studied populations of the barnacle, *Balanus glandula*, on the coast of Washington. Recruitment of larval barnacles to the rocky shore varied little from year to year, but the only young barnacles to survive were those that settled high in the intertidal zone. To test the hypothesis that mortality in the lower intertidal came primarily from predation by drilling whelks of the genus *Thais*, Connell fastened stainless steel wire cages over small areas on a pier piling (Fig. 4.4). Some control areas were left untouched; to others he affixed cages with roofs but no sides to simulate the physical environment of the cage while allowing access to predators. The results from the upper tidal level are summarized in Table 4.3 (the cages at lower tidal levels were less successful at

Figure 4.4 Predator-exclusion cages affixed to a concrete piling. The smaller barnacles are *Balanus glandula*. The larger are *Balanus cariosus*. The right side of the piling in the photograph is the south side, which is drier at low tide. This accounts for the left-to-right drop in the *B. cariosus* band (photograph by J. H. Connell).

Table 4.3 Survival of *Balanus glandula* with and without predators on the upper shore level of a concrete pier (Fig. 4.3). A dash indicates that no data were collected at that time (after Connell, 1970).

Treatment Predators + present − absent	Number of sites	Number of barnacles surviving at each of the following years of age										
		0.2	0.5	1.0	1.2	2	3	4	5	6	7	8
I. cages (−)	3	148	119	102	96	72	49	—	28	14	14	0*
II. cages (−)	3	243	—	134	—	70	—	20	8	7	0*	
III. open and roofed (+)	3	142	45	17	16	1	0					

* Predators gained access to the cages before these observations.

excluding *Thais*). No *B. glandula* survived past two years in the open; fourteen in cages survived seven years before *Thais* managed to enter and destroy them. The large barnacles in Fig. 4.4 are another species, *Balanus cariosus*, which is too large to be vulnerable to *Thais*. An invasion of the starfish, *Pisaster ochraceous*, subsequently destroyed the exposed individuals. Interestingly enough, *B. cariosus* under roofs but exposed to *Thais* survived the *Pisaster* onslaught and are probably 20–30 years old at the time of writing (Connell, personal communication).

Virnstein's study of soft-bottom infaunal communities in Chesapeake Bay gave even more dramatic results. In various cages sunk into the bottom he placed different densities of crabs and bottom-feeding fishes. The small clam, *Mulinia lateralis*, flourished in the absence of predation, increasing in abundance to enormous densities. Figure 4.5 shows the results of four months exclosure of predatory crabs and fish. The clam had reached phenomenal densities (Fig. 4.6). When crabs were accidentally allowed into an experimental exclosure containing one such dense population of *Mulinia* ($14\,700\ \mathrm{m}^{-2}$), they destroyed it in two months.

One might be tempted to conclude that only invertebrate ecologists are able to mount experimental studies spanning several generations. Such studies are certainly much more expensive with vertebrates but not impossible. Trautman *et al.* (1973) described a massive predator-removal project in South Dakota in the late 1960s, the object of which was to enhance the state's population of ring-necked pheasants. They established eight 100-square-mile study plots in eastern South Dakota. Although not identical in topography and vegetation, all plots were 'typical agricultural land'. On these eight plots they monitored densities of small mammals, pheasants, jackrabbits and cottontail rabbits and the mammalian carnivores, which preyed upon them. The carnivores of most interest were red foxes, raccoons, badgers, and spotted and striped skunks.

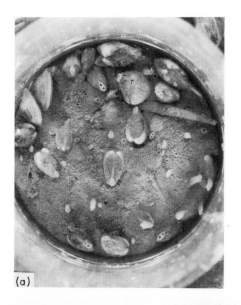

Figure 4.5 (a) An intact core from a predator-exclusion cage demonstrating the extraordinary densities achieved by *M. lateralis* when protected. (b) The same core with sediments and other animals removed (from Virnstein, 1977).

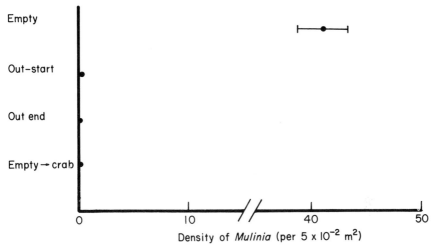

Figure 4.6   Density of M. *lateralis* in an experimental cage (labelled empty) after four months exclosure of crabs. 'Out start' and 'out end' are densities outside the cage at the beginning and end of this experiment. The treatment empty→crab refers to a cage with a large and growing population of *Mulinia* which was invaded by crabs halfway through the experiment. One was allowed to remain, and it destroyed all *Mulinia* (after Virnstein, 1977).

Since they thought the red fox was the dominant predator of pheasants, they targeted four experimental areas for the reduction of fox densities. One of the remaining four areas was left as an experimental control, and in the other three all small carnivores were removed by traps and poison. The results are summarized in Table 4.4. The experimental areas were not completely free of small carnivores, but the reduction was substantial. Of the various prey species, small mammals showed no response; jackrabbits increased enormously in the fox reduction areas and less in the areas where almost all carnivores had been removed; cottontail rabbits showed a response to small carnivore treatment but not to foxes alone; pheasants showed some reponse to reduction of foxes alone and a substantial increase in response to the reduction of all species of small carnivores.

This study is, without question, the largest, best documented, and most expensive predator-removal project in wildlife annals. The data are interesting from a variety of perspectives. For example, the possibility exists that competitive relationships among some of the prey species were altered by the changed predatory environment. That the data are not more available to ecologists is unfortunate.

What do these studies tell us? We become convinced that if predators are removed the prey respond, but is that enough? The arguments in the second and third chapters lead one to expect that altering a predator–prey system by reducing predators could have consequences for both the equilibrium

Table 4.4 Results of the predator-removal project in South Dakota. Data are the percent changes in population density (after Trautman et al., 1973).

Species	Fox reduction areas Reduction areas %	Control areas %
Fox	− 83	− 50
Pheasants	− 39	− 58
Small mammals	+ 66	+ 51
Jack rabbits	+ 326	+ 190
Cottontails	+ 62	+ 80
	Small carnivore reduction areas	
Pheasants	+ 75	− 19
Small mammals	− 7	− 26
Jack rabbits	+ 249	+ 219
Cottontails	+ 133	+ 83
Fox	− 78	− 44
Raccoon	− 54	+ 41
Badger	− 68	+ 57
Skunks	− 64	+ 23

abundance and the stability of the prey. The necessarily cautious approach is to assume that neither influence is a guide to the other. To be extra careful, one should also allow for the possibility of oscillations arising as a result of such manipulations. Unfortunately, with the exception of Huffaker and Kennett (1956), none of these studies were so interpreted. In all cases where dynamical predictions were made, the stability properties of the perturbed systems were assumed to be the same as the systems with predators present. When one bases an experiment upon the limiting-factor type of question, 'do predators have an effect upon prey numbers', one is asking too simple a question. The consequence is that one wildlife biologist, Errington, could focus entirely upon the effects of predators on the stability of prey numbers and ignore their influence on equilibrium densities, while others are equally myopic in precisely the opposite way.

Is this a trivial point, of interest only to theorists? Trautman and his coworkers concluded their report with the recommendation that a large-scale popular effort be mounted to reduce all small carnivores in South Dakota. The secondary economic benefits from twice as many pheasants would more than pay the costs, they claimed. But their data can be interpreted another way. Equally consistent with the experimental results is the conclusion that the average or equilibrium density of pheasants was unchanged by the removal of predators; what changed was the stability of the pheasant population. The South Dakota workers may have observed only the increase phase of a now oscillatory pheasant population, an increase which would be followed in a few years by a collapse. The benefit to the South Dakota sporting industry might, as a consequence, be followed quickly by financial disaster.

The problem lies not in the proper interpretation of predator-removal experiments, doomed to be a frustrating task; the problem lies in the original experimental design. Unless the field biologist follows his population for a long time, he will find it difficult if not impossible to separate out stabilizing effects from effects upon prey equilibria. But there are alternative experimental approaches in the literature, and these are worth exploring. Two come to mind, each quite different from the other. The first was employed by Bergerud (1971) in his studies of the interaction of lynx and caribou.[4]

Bergerud ran across a curious predatory interaction in the course of studying caribou on Newfoundland. While tracking population fluctuations of the Newfoundland herds, he discovered that, apart from hunting mortality, population growth and decline depends primarily upon variation in the survival of calves from birth to six months of age. Figure 4.7 displays the estimated annual rate of increase as a function of calf mortality. Bergerud's

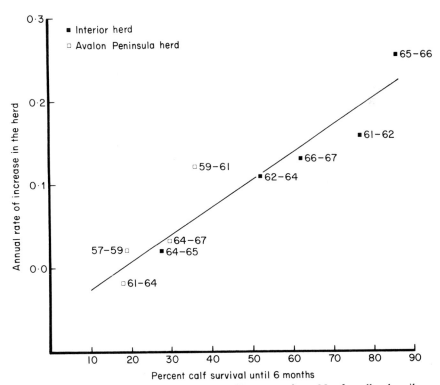

Figure 4.7   Dependence of the annual rate of increase of two Newfoundland caribou herds upon the percentage of calves surviving to 6 months of age (after Bergerud, 1971).

search for the remains of dead calves yielded 114, of which 84 had been bitten by lynx. When one thinks of predators of caribou, one thinks typically of wolves, not lynx. But wolves are absent from Newfoundland. Lynx, in turn, are traditionally associated with snowshoe hare. Newfoundland lynx rely on snowshoe hare for food, certainly, but they also apparently ambush wandering caribou calves in the summer months.

The circumstantial evidence for the importance of lynx predation was good at that point, but Bergerud went one step farther, designing an experiment to test his hypothesis. One calving ground, Middle Ridge, was designated as a lynx-removal area. The Pot Hill calving ground served as a control. Forty-four lynx were trapped from Middle Ridge early in 1965. Table 4.5 displays the response of the caribou herd and reveals that calf survival was substantially greater at Middle Ridge.

Table 4.5 The percent survival of calves in the Middle Ridge experimental calving ground and the Pot Hill control calving ground in 1965 (after Bergerud, 1971).

	Middle Ridge %	Pot Hill %
Calf survival to June (about 2 months of age)	85	49
Calf survival to October (about 6 months of age)	16.3	2.7

The value of Bergerud's study lies more in his approach than in the specific result. The separation of total mortality into that affecting different age classes is a well-known technique called key-factor analysis.[5] The important feature of this technique is that it reveals something about population regulation apart from changes in equilibrial abundance. In this sense the Newfoundland study is unique in the literature of wildlife biology.

The second study, reported by Slade and Balph (1974), dealt with a population of Uinta ground squirrels observed near the Utah State University Forestry Field Station from 1964–1971. The primary habitat for this population was a lawn area of 0.9 ha surrounding the building. From 1964 to 1968 the number of ground squirrels on the lawn averaged about 50. In 1968 and continuing through 1971 the density was reduced and maintained by trapping to an average of about 33 animals. A traditionally heavy source of mortality for this population was predation by badgers during the winter. With the reduction in ground squirrel density, badger predation dropped off. The mean number of badger diggings per hibernating squirrel before the reduction was 0.46; the mean number after was 0.22. Losses during the winter decreased 14%. Apparently the badgers chose to hunt elsewhere.

The appeal of this study lies not in the simultaneous controls nor in the numbers of replicates; there were none. Its appeal is in its design and interpretation. The idea that one could investigate the influence of predation by altering *prey* densities seems not to have occurred to most workers in this field; yet it is precisely this design which was advocated by Nicholson (1957) and Murdoch (1970) as the most direct way of detecting population regulation. When done properly, with simultaneous controls, it gets around the assumption of constant equilibria. And in interpreting their data, Slade and Balph did not agonize over the single limiting factor which must account for all of the Uinta ground squirrel's dynamics. They acknowledged that both predation and dispersal, motivated by intraspecific aggression, must play critical roles; moreover, they suggest that the two factors interact to form a complex that is too interrelated to be treated as merely the additive or multiplicative effects of predation and dispersal considered alone.

A few simple conclusions follow from perusal of these case histories. The first is that experimental studies potentially reveal so much more about dynamic relationships than descriptive studies do that little justification exists any longer for doing the latter, unless what is described is a natural experiment. The second point is that simple predator-removal experiments must be followed for a long time to try to discriminate the separate experimental effects on equilibria and stability. This is a difficult separation to make, at best, and no predator-removal studies to date have done it successfully. Thirdly, a specific focus on population regulation either by the narrowing down of regulatory mortality to one age-class or by the manipulation of prey densities is probably a more productive experimental approach.

The manipulative experiment I would like to see performed would consist of a factorial arrangement of enhanced, diminished and control prey populations compounded with enhanced, diminished and control predator populations. We are unlikely to see such a study done with vertebrates; the expense and difficulty would be enormous. But the design is not impractical for some invertebrate systems. Ultimately the best approach is to develop a good model of the system under investigation and conduct the experiments suggested by the predictions of that model.

NOTES

1. Wolfe and Allen (1973), Peterson (1976, 1979), Peterson and Allen (1974), Peterson and Page (in press).
2. These and many other examples of successful regulation of pests by predators and parasites can be found either in the encyclopedic volume by Clausen *et al.* (1977) or in a less-detailed but more accessible article by Caltagirone (1981).
3. There are several older studies which for various reasons yield ambiguous

results and will not be discussed. See Edminster (1939), Robeson *et al.* (1949), and NY State Conservation Department (1951).

4. He has also provided a simplified digest of his work for the popular audience (Bergerud, 1983).

5. This is described in some detail in Varley *et al.* (1973). One should be cautious in the use of this method. It is predicated upon the assumptions that age-related mortality factors are independent and additive in their effect, assumptions which may be unjustified in any particular case.

5 Self-limitation of prey and predator populations

Whatever else may be said of predation,
it does draw attention.

Errington (1946)

This and the subsequent four chapters explore a variety of complications which seem to pop up frequently when dealing with real predator–prey systems. The goal in these chapters is to discover if a specific feature of the natural history of a predator–prey interaction will have the same effect, wherever encountered. The topics treated do not, by any means, exhaust the list of reasonable complications which occur in predator–prey interactions. Beyond their prevalence, they share the redeeming feature that their contributions to population dynamics are beginning to be understood. For some of these features we have good, well-focused experiments, for others suggestive field data, and I shall try to complement existing empirical data with a simple account of the pertinent theory.

Relevant to this, it might be worthwhile to begin by discussing a common misconception among empirically-oriented ecologists. One could easily infer from Chapter 2 that models exist in either of the two forms discussed and no other. The first form is exemplified by Equations (2.7) and (2.8), simple models containing only the most necessary and ubiquitous features of predation; the second type is characterized by the three wildlife models, an approach dominated both by concern with a specific community and by the need for predictive accuracy. But in fact these two classes of models merely represent the poles of a spectrum of theoretical approaches. Much current research in predation theory consists of an intermediate approach in which a simple model is complicated in various ways and the effects of these changes examined. The tactic is to develop a mathematical description of a sub-process, searching for example or digestion, imbed it in a predator–prey model, and then ask questions: is an unstable model stabilized; is a stable model destabilized; is a constant equilibrium made cyclic? Those who do this kind of research realize that models so modified are less general than their precursors but accept this as a fair price to pay for a concomitant increase in relevance and predictive power. Success in the use of this research tactic requires recognition that inevitably, in the course of complicating things, one will arrive at a model which is not only inapplicable to all but a small subset of predator–prey systems but is so awkward mathematically that it may be nearly as difficult to understand as the natural system it was meant to mimic.

I shall deal first with the effects of environmentally-imposed limits on the abundance of prey. Evidence which suggests that prey are near the limits of their food supply or other resource base has been widely taken to mean that predation is unimportant. Errington provided an example of such reasoning in the interpretation of his studies of muskrats (Chapter 3). From his observation of wandering animals covered with wounds from fights, he inferred that predation by mink was of secondary importance to intraspecific competition; if mink were significantly depressing muskrat numbers, the surviving individuals would be uncrowded and disinclined to fight with one another. Kruuk's study of the spotted hyena in Africa employed the same kind of logic (Kruuk, 1972). He noticed that wildebeest in the Serengeti showed direct and indirect evidence of starvation; wildebeest in the Ngorongoro Crater, by comparison, were well fed. His conclusion: hyenas were not regulating wildebeest on the Serengeti Plains but were in Ngorongoro Crater. But, as is the case with most field studies, we find that the effects of predation upon the equilibrium population size are confounded with its effects on population regulation. Neither Errington nor Kruuk offered any direct evidence for the presence or absence of predation as an influence in the regulation of prey species.

The question of the effects of external limits to prey abundance has been addressed by a variety of experimental studies. The best of these have employed microorganisms in the laboratory. Luckinbill (1973, 1974) re-examined the dynamics of the ciliate protozoan *Paramecium aurelia* and its ciliate predator, *Didinium nasutum*, two species that had been used by Gause (1934) in one of the first experimental investigations of predation. Gause's classical studies established that the interaction was unstable, a result subsequently confirmed in countless college biology classes. But Luckinbill discovered that if the culture conditions were changed both by the addition of methyl cellulose, a chemical which increases the viscosity of the culture medium, and by dilution of the nutrient medium, then one could get persistence of the predator and prey. High nutrient systems suffered the more traditional fate of elimination of one species or the other (Fig. 5.1).

Veilleux (1979) repeated Luckinbill's experiments, with comparable results, and continued work to estimate the shapes of the predator and prey isoclines for each of seven different concentrations of the nutrient medium (Fig. 5.2). Were one to apply the simple theory to these results, one would expect a stable equilibrium point at a Cerophyll concentration of $0.68 \, \text{g} \, \text{l}^{-1}$ to gradually change to a limit cycle of increasing amplitude as the strength of the medium increased. Stable limit cycles of increasing amplitude were in fact observed, but a stable equilibrium point was not.

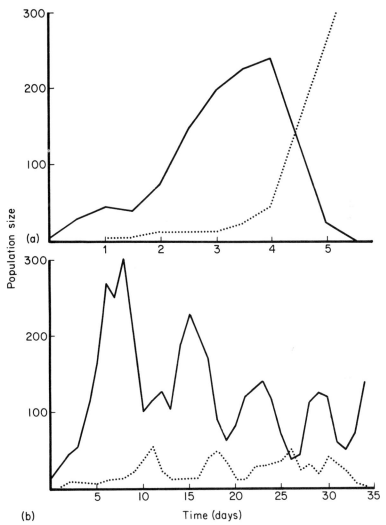

Figure 5.1   Paramecium numbers (solid lines) and *Didinium* numbers (dotted lines) in (a) a rich medium showing persistence for only 5.5 days and (b) a dilute medium showing persistence for 34 days (after Luckinbill, 1974).

Veilleux suggested that a time lag in his system violated the assumption of instant feedback in Equations (2.7) and (2.8). Others have demonstrated the destabilizing potential of time lags (Wangersky and Cunningham, 1957; Maly, 1969). Several candidates for such a lag exist. The most obvious is the fact that the experimental culture of these protozoa was not continuous. Veilleux grew them in a Petri dish in 6 ml of medium, one-half of which was replaced every two days.

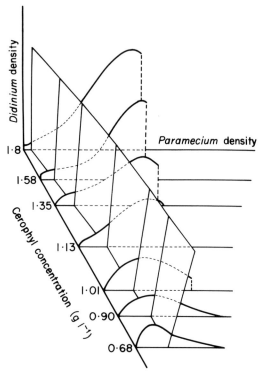

Figure 5.2   Dependence of the phase-plane portrait upon the strength of the medium in the *Paramecium–Didinium* interaction (from Veilleux, 1979).

Re-examination of Figs 2.8 and 2.9 should suffice to reveal what the simple theory has to say on this point. Note that the stable equilibria in the various forms of the general model are all near the self-imposed limits on prey numbers, suggesting that such limits tend to stabilize systems. The reasoning behind this assertion can be presented mathematically, but the argument is not complex; if the equilibrium density of prey in the presence of predators is only a little lower than the equilibrium in the absence of predators, then a rise in prey density above the lower equilibrium will be met not only with a rise in predation but with an internally imposed reduction in population growth rate. The result will be a tendency to return to the steady state. Further exploring this theoretical relationship between prey limits and stability, Rosenzweig (1971) posed the 'paradox of enrichment'. The idea was that enrichment of a system, by raising the limits on prey density, reduced the internal contribution to stability. In Fig. 5.3 such a hypothetical shift in richness is graphed on the phase plane. The relative position of the predator isocline is shifted to the left; a population initially at a stable equilibrium is now cycling. Paradoxically, if the prey population is experiencing lower fecundity and greater mortality as a

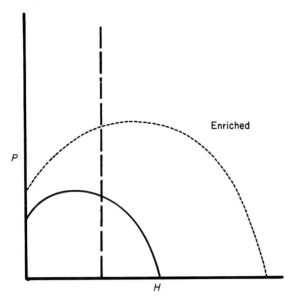

Figure 5.3   Alteration of the isocline of the prey by enrichment of its food supply. Note that the equilibrium point is shifted from the right to the left of the maximum of the prey curve and is thereby rendered unstable.

result of lack of certain resources, provision of these resources may produce the opposite effect from that intended: extinction rather than increase.

An alternative theoretical portrait of the influence of environmental limits on prey abundance emerged from a study by Tsuchiya *et al.* (1972) on the ameboid stage of the slime mold, *Dictyostelium discoideum*, feeding upon the bacterium, *Escherichia coli* in continuous chemostat cultures. Chemostats are sealed culture vessels with two openings. Through one, a nutrient medium is fed continuously into the culture; through the other, sufficient fluid is withdrawn to maintain the volume. Such an apparatus is difficult to maintain uncontaminated for any length of time, but it has proved valuable for several reasons. Among these is the fact that it meets nicely one of the basic assumptions of models employing differential equations: the input of energy or nutrients in the form of fresh medium is continuous in time, not discrete.

The operator of a chemostat has control over two features of the medium fed into the culture, its concentration and its rate of flow. The rate of feed is typically expressed in the inverse form: the holding time. This is the ratio of the culture volume to the rate of feed and measures the average length of time a quantity of fluid remains in the vessel. In a homogeneous culture, holding time also measures the average time a cell in that volume of fluid can expect to remain in the chemostat before it is washed out.

Tsuchiya *et al.* (1972) developed a model of this interaction which took explicit notice of the concentration of the limiting nutrient, in this case glucose.

If one defines C_B and C_A to be the concentrations of bacteria and amoebae in the medium and C_G and C_{GF} to be the concentrations of glucose in the medium and in the feedstock and if, further, one defines θ to be the holding time and μ_A and μ_B to be the maximum growth rates of the amoeba and the bacterium, then the three equations of the model are

$$\frac{dC_G}{dt} = \frac{C_{GF}-C_G}{\theta} - a_G\left[\mu_B\frac{C_G}{K_G+C_G}\right]C_B \tag{5.1}$$

$$\frac{dC_B}{dt} = \left[\mu_B\frac{C_G}{K_G+C_G}\right]C_B - a_B\left[\mu_A\frac{C_B}{K_B+C_B}\right]C_A - \frac{C_B}{\theta} \tag{5.2}$$

$$\frac{dC_A}{dt} = \left[\mu_A\frac{C_B}{K_B+C_B}\right]C_A - \frac{C_A}{\theta} \tag{5.3}$$

in which a_G is the uptake of glucose per unit bacterial growth and a_B is the consumption of bacteria per unit of amoeboid growth. The parameters K_G and K_B are saturation terms for the uptake of glucose and the consumption of bacteria. The equations are straightforward and noteworthy primarily for their use of the Monod function in parentheses to describe limits to rates of uptake and predation (Monod, 1942). Fig. 5.4 compares the predictions of the model to the actual behavior of the chemostat. The fit is quite good until the third week. Apparently all cultures began to depart from the model after three to five weeks. The reasons for this were not clear but probably represented long-term physiological changes rather than genetic changes.

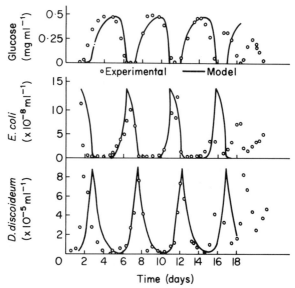

Figure 5.4   Comparison of experimental data (open circles) with the predictions of Equations (5.1)–(5.3) (solid lines) (from Tsuchiya et al., 1972).

The correspondence between model and reality is good enough to justify examining what Equations (5.1)–(5.3) have to say about the effects of increased richness of the medium. Figure 5.5 displays different classes of predicted population dynamics on the parameter plane formed by plotting holding time, θ, versus glucose concentration in the feed, C_{GF}. If holding time drops below the minimum time a cell must have to grow and divide, then the population must wash out regardless of the availability of nutrients. Likewise,

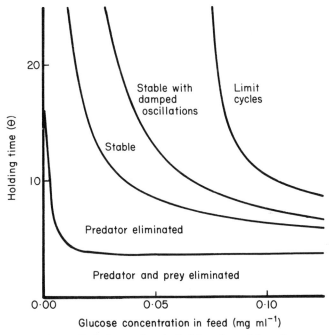

Figure 5.5   The effects of jointly varying the holding time of the culture vessel and the glucose concentration of the feed upon the dynamics of Equations (5.1)–(5.3) (after Tsuchiya *et al.*, 1972).

nutrient solutions so dilute as to barely allow cells to divide can only permit maintenance of a population when the holding time is quite long. By interpolation, therefore, one should expect holding time and nutrient substrate concentration to interact roughly inversely in the setting of the minimum conditions for the survival of the chemostat culture, and this seems to be the case.

What is less intuitively obvious is that, for tolerable holding times, increasing the glucose concentration takes a culture from a stable equilibrium, to which the populations approach asymptotically, to a stable equilibrium approached in an oscillatory fashion and finally to a limit cycle. This alternative theory, both different in structure and well supported by experimental evidence, serves to confirm the prediction of the basic theory, that tight environmental limits on prey numbers promotes stability.

Remarkably little evidence exists to counter this hypothesis. The only such result of which I am aware was obtained by Drake (1975), again using *Dictyostelium* preying upon *E. coli* in a chemostat. Under culture conditions producing oscillations, Drake found that enrichment of the medium with peptone reduced the amplitude of the oscillations and increased their period. Peptone is not equivalent to glucose, the normal limiting substrate, and could directly benefit the amoebae, but the fact remains that the procedure did simultaneously increase bacterial growth and stability.

The intraspecific limitation of predators by their own social interactions is substantially more difficult to investigate experimentally than is the self-limitation of prey. The behaviors, usually aggressive, which limit the numbers of predators are frequently inflexible for a particular species, and this makes difficult the conduct of experiments in which these behaviors must be altered. Although one predatory species can be compared to another, changes in intraspecific aggressive behavior are usually accompanied by a constellation of other changes. And which of these is actually responsible for an observed change in dynamics can be difficult to discover.

Nonetheless, such phenomena as territoriality, dominance, or even irritability are common and undoubtedly influence population dynamics. Among wolves, for example, an intricate social hierarchy results in a predation rate per pack which is fairly constant and insensitive to group size (Haber, 1977). Those wolves which direct the hunt are also the first to feed at a fresh kill. The onset of the next hunt must await the renewal of the hunger of these well-fed dominant animals, whether or not the subordinate animals are starving.

Much of the evidence on this point is merely observational and can give rise to misinterpretation, just as do limits to prey numbers. Hornocker (1969, 1970) documented predation upon mule deer and elk by mountain lions in Idaho over a four-year period. During this time both deer and elk populations increased, amid signs of overbrowsing on winter forage. Mountain lion numbers changed very little during this time, apparently because the animals are antisocial and young adults dispersed out of the area. Hornocker simultaneously concluded that lions were stabilizing the prey but were not the 'limiting factor', his definition of this term being that factor which most significantly reduces a population's rate of increase. This confusing conclusion emerged from an attempt to communicate that although lions did not significantly reduce the equilibrium abundances of mule deer and elk they may have caused those equilibria to become more stable. The latter result could not be demonstrated with a study lasting only four years, of course, and so must remain conjectural.

Insect ecologists have been responsible for a great deal of what we know about the results of predators interfering with one another (Hassell, 1978;

Beddington *et al.*, 1978). A data set typical of this class of studies comes from Huffaker's and Matsumoto's (1982) work on the parasitic behavior of the ichneumon wasp, *Venturia canescens*. They allowed single wasps and groups of ten to search for and parasitize late-instar larvae of the moth *Anagasta kühniella* in laboratory arenas for 24 hours. Figure 5.6 reveals that at four different densities of host larvae the numbers of ovipositions per adult parasitoid are substantially reduced. The source of this reduction lies mostly in

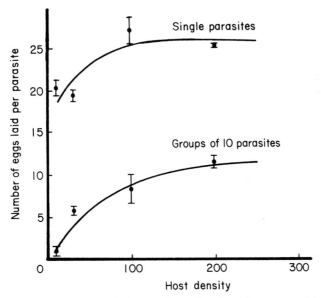

Figure 5.6 The dependence of the number of ovipositions per adult *Venturia canescens* upon the density of host larvae both when hunting singly and in groups of ten wasps (after Huffaker and Matsumoto, 1982).

the interference of parasitoids with one another during the process of searching (Hassell, 1978). Obviously a behavioral process such as this is not restricted to wasps. Salt and Willard (1971) discovered the same effect in Forster's Terns hunting small fish in salt flats on the San Francisco Bay. As the numbers of birds increased, the rate of predation per bird declined. A larger proportion of each bird's time was taken up in the avoidance of collisions, and less could be spent searching. Using quite different organisms, Salt found the same pattern in two protozoan predators of *Paramecium*; both *Didinium* and *Woodruffia* show a decline in individual predatory efficiency at high densities of their own kind (Salt, 1967, 1974).

The basic theory of Chapter 2 suggests that predator interference is strongly stabilizing. One can get a feeling for this result by examining Fig. 2.8; the predator isocline there includes an abrupt upper limit on population growth, independent of food supply, and generates an asymptotically stable equilibrium. The effects of such interference have been addressed in the most detail with a difference-equation model for parasitism originated by Nicholson and Bailey (1935) in which the numbers of hosts and parasitoids at time $t + 1$ follow from

$$H_{t+1} = \lambda H_t \exp(-aP_t) \tag{5.4}$$

$$P_{t+1} = H_t[1 - \exp(-aP_t)] \tag{5.5}$$

in which λ is the finite rate of increase for the host population and a is the rate of parasitism per parasitoid. The forms of these equations are easy to understand if one accepts three assumptions. First, hosts and parasitoids are assumed to reproduce only at discrete points in time, and reproductive performance is not inhibited by crowding. Second, in the intervals between these points, parasitoids constantly reduce the numbers of hosts. The reduction follows from a process of random search and is easy to visualize. If H is the number of hosts present per unit area and if the rate of search per parasitoid is constant at a, then in a short interval of time $d\tau$, the number of hosts discovered should be $aPH \, d\tau$. The discovery of these hosts reduces the density of the remaining individuals, such that

$$-dH = aPH \, d\tau \tag{5.6}$$

Integrating this subject to the condition that at $\tau = 0$, $H = H_t$, we find that

$$H_{t+\tau} = H_t \exp(-aP\tau) \tag{5.7}$$

At time $\tau = 1$, each surviving host in Equation (5.7) produces λ offspring to generate Equation (5.4). And the third premise is that those hosts which do not survive each produce one parasitoid for the next generation; hence the form of Equation (5.5).

The Nicholson–Bailey equations are unstable for all values of the constants a and λ. But not much weight should be attached to this quality since the equations embody a number of simplistic assumptions, among them that the rate of parasitism per parasitoid is constant and independent of parasitoid density. Hassell and May (1974) have analyzed what happens to population stability when the constant a declines with increasing parasitoid numbers according to the function

$$a = QP_t^{-m} \tag{5.8}$$

The constant m mediates the depressing influence of parasitoid density and is known as the interference constant. Figure 5.7 shows that although the original Nicholson–Bailey model $(m=0)$ is unstable it can be stabilized by allowing m to assume some values between zero and one. One could argue that Equation (5.8) is merely curve fitting and that it has no biological roots, but

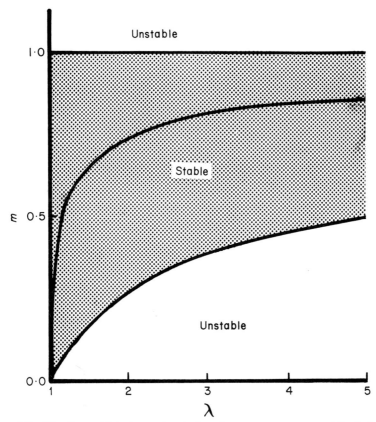

Figure 5.7   The effects of jointly varying the interference constant, *m*, and the finite rate of increase, *λ*, upon the dynamics of the Nicholson–Bailey model, modified by Equation (5.8) (after Hassell and May, 1973).

Hassell (1978) points out that behaviorally explicit models of interference give much the same result.

   In summary, it appears that the limitation of prey numbers by agencies other than predation and the limitation of predator densities by mechanisms not related to food supply both contribute measurably to the stabilization of exploitative interactions.

6 Age and size structure in predator and prey populations

The old lion perisheth for lack of prey . . .

Job 4:11

For the sake of convenience we commonly treat populations as though they were homogeneous units, comprised of genetically identical, even-aged organisms having indistinguishable developmental histories. Since such a practice is, with few exceptions, an offense against nature, we ought to devote attention to the effects of relaxing this assumption. The influence of genetic variation upon the predatory process is taken up in Chapter 11. Here I address the effects of variation in age and size in both predator and prey populations.

A well-developed body of basic theory exists to deal with the influence of age structure upon the dynamics of a single species.[1] But this theory has traditionally avoided circumstances where age-specific mortality and fecundity schedules must vary, either with the density of the population in question or with the densities of the other species with which that population interacts in the ecological community. Populations involved in predatory or competitive interactions exhibit mortality and fecundity schedules which change continually in a determined fashion. The results of such age-structured interactions can be surprising and counter-intuitive. The literature on zooplankton, for example, contained for a number of years the hypothesis that large cladocera were competitively dominant to small cladocera because their broader diets and greater filtering rates gave them superior feeding efficiency (Brooks and Dodson, 1965). In laboratory studies of competition Neill (1975) found the opposite to be true; two small cladocerans dominated larger crustacean species. The explanation of this result lay in the age structure of the competitors. Large individuals do not originate as such, of course; they must develop from small larvae. The smaller species were sufficiently more efficient than the early stages of the larger species that they formed a competitive bottleneck to recruitment of large adults. Ignoring age structure in this circumstance proved to be unwise. Given the dynamic richness of predation, by comparison to competition, we should not be surprised to discover that age and size structure in predator and prey populations could produce an even greater variety of surprising results.

The first step in looking at this type of complication is to establish that the predatory process is, in fact, sensitive to age and size variation in the two populations. Then we need to know if such variation as exists is important enough to worry about. Would it be acceptable, for example, to gloss over some of it by computing an average rate of predation per predator? The answer, as we shall see, depends upon the temporal constancy of the structure of the populations. In particular we need to know if predation itself effects systematic qualitative changes in the age and size composition of the interacting populations.

The two most well-known, age-dependent features of predation are the development of hunting skills and the relative vulnerability of very young and old prey. The ineptness of young predators is well documented.[2] The development of facility in prey capture frequently requires a long and tedious apprenticeship, especially among birds and mammals. The oystercatcher, to use a well-documented example, takes up to 43 weeks to learn either of the two common methods for opening and consuming mussels (Norton-Griffiths, 1967). Young oystercatchers probably require three to four years to perfect their technique; at least they defer breeding until that age. Young invertebrates also hunt less effectively, but here the cause lies more in their inadequate strength and size than in the learning of motor skills and foraging tactics. Thompson (1975) compared the predatory behavior of different nymphal stages of the damselfly, *Ischnura elegans*, upon a range of sizes of the cladoceran, *Daphnia magna*. Figure 6.1 contains graphs of both the rate of predation and the handling time per prey item as functions of predator and prey size. Small *I. elegans* nymphs have difficulty coping with larger *Daphnia*; the smallest nymphs caught so few of these that Thompson was unable to estimate their handling times. This is a recurrent pattern among invertebrates. The upper bounds on body size, imposed by their basic morphology and physiology, tend to result in predators more similar to prey in size and strength than is common among vertebrates. It would be reasonable, in summary, to recognize that a predatory population may contain individuals of widely varying competence.

The idea that young and old prey are more vulnerable to predation remains a very firmly-held belief among biologists. A large body of observational data supports the concept (Mech, 1970), although it is best to recognize that peculiarities of natural history can sometimes alter the pattern. For example, Wasserzug and Sperry (1977) discovered that the hylid frog, *Pseudacris triseriata*, is most vulnerable to predation by garter snakes neither as tadpole nor as adult, but during the transformation from the one form to the other. Apparently transforming frogs neither swim as fast as tadpoles nor jump as well as adults. But such stories are the exception by which we prove the rule that predation falls disproportionately on the very young and very old. The evolutionary processes underlying this pattern will be explored in Chapter 11, but I shall anticipate here one of its implications: the heaviest predation

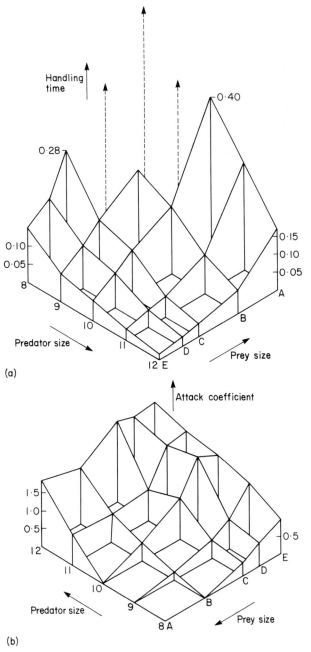

Figure 6.1 The influence of predator and prey size on (a) the amount of time spent handling cladoceran prey by nymphal damselflies, *Ischnura elegans* (very small damselflies cannot eat large cladocera), and (b) the rate at which *I. elegans* of different sizes attack variously sized prey (from Thompson, 1975).

mortality usually occurs in the subset of the prey population least important to future reproduction. Old animals have little or no future reproductive contribution to make; young animals can be quickly replaced.

What are the dynamic consequences of these and other age- and size-related variations in the interaction? Perhaps the only defensible answer is that we are only beginning to understand certain special cases and that any sweeping generalizations that might be offered are unlikely to be correct. Certainly an inverse correlation of age-dependent vulnerability with relative reproductive value would lead one to expect predation to exert a disproportionately small influence. A general argument bearing on this relationship was offered by Rosenzweig (1978), in which he attempted to sketch out a phase-plane portrait of the shape of the prey isocline in the circumstance where predation is compensatory – that is where subordinate prey at high densities become more vulnerable to predation. Figure 6.2 displays Rosenzweig's proposed modification. He suggested that a predator selectively killing animals with lesser reproductive competence exerts a relatively smaller impact upon prey numbers. The consequence is an elevated H isocline as H gets large. Rosenzweig also suggested that the P isocline would shift to the left. The

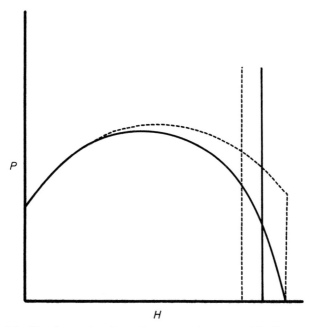

Figure 6.2   Modification to the phase-plane portrait suggested by Rosenzweig (1978) to treat predators which selectively remove prey of lesser reproductive value. The solid line represents a non-selective predator; the dashed line a selective predator. The result would be a major increase in the numbers of predators, a minor decrease in the numbers of prey, and a small decrease in the stability of the interaction.

dynamic consequences would be a marginal decrease in the local stability of the equilibrium point and probably a substantial reduction in the likelihood of stochastic extinction of the predators, by virtue of their substantially higher equilibrium numbers.

Rosenzweig's conclusions should be restricted to those circumstances in which knowledge of H alone allows one to predict exactly the number of vulnerable prey. His graphical tactics are risky if the number of vulnerable prey is a less precisely-specified variable. In that case, the trick of redrawing the isoclines would amount to projecting on to two dimensions the three-dimensional phase space with axes consisting of densities of predators, normal prey, and especially vulnerable prey of low reproductive value. Such a projection allows for graphical stability analysis only if one assumes that the qualitative dynamical behaviors of models of dimension three do not exceed in complexity those of dimension two, an assumption which can be spectacularly unfounded.

Smith and Mead (1974) examined deterministic and stochastic versions of a less general model in which prey were divided into two age classes of different vulnerability. They concluded that, while age structure could either enhance or reduce stability depending upon how it is expressed, normally the smaller the proportion of the life cycle vulnerable to predation, the more stabilization resulted.

Several papers have been published on the circumstance in which predator populations have an age structure and prey do not. Maynard Smith and Slatkin (1973), Hastings and Wollkind (1982), and Wollkind et al. (1982) developed continuous-time models which predict that stability is not enhanced by age-structured predators. Age structure operates as a time lag in the system. Beddington and Free (1976) examined a discrete-time model and found age structure to put less of a damper on stability, perhaps because a time lag was already present in their model.

A strong argument can be made that highly age-specific predation should induce oscillations in the two populations. Age-structured populations are normally vulnerable to periodic environments. This is easily illustrated by use of a physical metaphor employed by Auslander et al. (1974). Conceive a population as a collection of sand grains on the continuously-moving porous conveyor belt in Fig. 6.3. 'Births' originate from the hopper, which drops sand onto the belt at point 0. As the sand is carried along, a portion of it sifts through the belt, mimicking death. At a distance α from the hopper, the belt passes over a scale. The amount of sand on the scale represents the number of individuals of reproductive age and, as such, sets the rate of flow from the hopper. This physical analog easily allows one to envision the results of an episode of age-specific predation. Imagine the removal of all the sand from a short length of the belt between 0 and α. As that portion of the belt moved over the scale the valve on the hopper would close down and produce a dip in the level of the sand. In this fashion a traveling wave would be established which might not

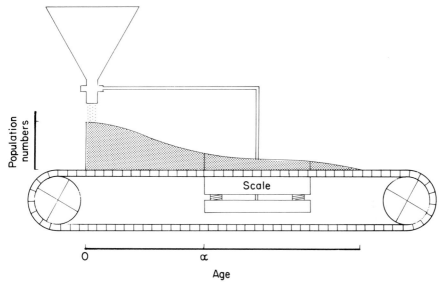

Figure 6.3   A population as a collection of sand grains on a continuously-moving and porous conveyor belt.

disappear for a number of generations. Obviously such populations are most sensitive to age-specific disturbances at certain frequencies (Oster and Takahashi, 1974).

Things get complicated when two age-structured populations interact in an age-specific fashion; traveling waves in the age structure of one population will induce waves in the other, and these in turn can feed back into the first population. The nature of these patterns could be expected to depend upon the generation times of the species, the degree of age dependence of the interaction, and a variety of other factors.

I shall illustrate how this can operate with the empirical study which motivated Auslander *et al.* This dealt with the Mediterranean flour moth, *Anagasta kühniella*, and its predators and parasitoids (White and Huffaker, 1969a,b). Moths of this species kept on a rolled wheat medium fluctuate in numbers of adults according to the pattern in Fig. 6.4. *Blattisocius tarsalis*, an egg-eating mite, when cultured with *Anagasta* produced the curious cyclic pattern in Fig. 6.5. The similarity of the two replicate systems suggest that the pattern is a property of the interaction and not merely stochastic fluctuation. A second series of cultures with the larval parasitoid, *Venturia canescens*, also exhibited regular fluctuations (Fig. 6.6), although these differed from those produced by *Blattisocius*. The original force behind such long-term cyclic behaviors is not obvious. It could be environmental fluctuations extrinsic to the predator–prey system; such fluctuations need not be enormous, merely coincident with a natural period of the system. Auslander *et al.* presented an

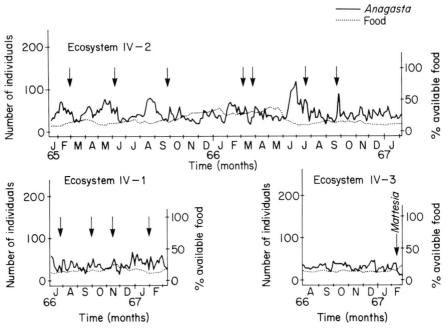

Figure 6.4 Population numbers of dead adult *Anagasta kühniella* collected twice weekly from replicate cultures of the moth. The replicate labelled 'Ecosystem IV-3' suffered an invasion of a protozoan parasite, *Mattesia* sp., and was terminated shortly thereafter. Arrows indicate times of application of acaricides necessary to keep the cultures free of mites (from White and Huffaker, 1969a).

alternative portrait. Their model of the moth-wasp system generated two classes of stable equilibria, one in which all age groups are represented and another featuring traveling waves of uniformly-aged cohorts of hosts and parasitoids. Such cycles are endogenous, emerging solely from the interaction of the host and its parasitoid.

To overlay variation in size on this complexity would seem to be inviting chaos, but in fact we can make a few tentative generalizations about the effects of developmental change and plasticity. The most well-known system in which this has proved important is Gause's study of the predatory interaction of *Paramecium* and *Didinium*, mentioned in the last chapter. Those experiments revealed a peculiar property of the predator. *Didinium* responds to starvation stress in two ways: some individuals form resting stages or cysts which sink to the bottom and no longer influence the system, but others continue to divide into smaller cells. These move faster in an apparent attempt to leave the immediate vicinity and migrate to areas with more prey (Salt, 1979). In the laboratory, where emigration is constrained, this response accelerates predation upon already depleted *Paramecium* populations and drives them to extinction. As unusual as this behavior may seem to be, its consequences for

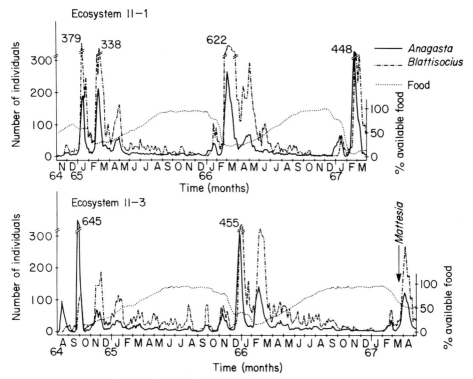

Figure 6.5 Replicate cultures of the Mediterranean flour moth and *Blattisocius tarsalis*, an egg-eating mite. The moth population (solid line) is the number of dead adults at the semiweekly census. Numbers of mites (dashed line) are those visible on the inside surfaces of the experimental chamber. Ecosystem II-3 was terminated shortly after *Mattesia* infected the moth population (after White and Huffaker, 1969a).

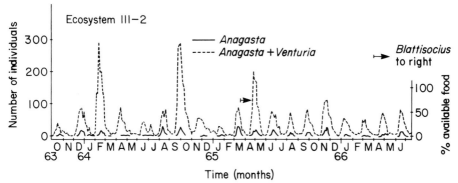

Figure 6.6 Population fluctuations of *A. kühniella* and its ichneumon parasitoid, *Venturia canescens*. Solid lines are the numbers of dead *A. kühniella* removed at the semiweekly census. The dashed line represents the sum of the *A. kühniella* and the adult *V. canescens* present. The mite, *B. tarsalis*, infected the cultures at the point indicated by the arrow (after White and Huffaker, 1969b).

population dynamics are obvious; it is strongly destabilizing. This was made clear both by mathematical analysis (Gause *et al.*, 1936) and by comparing Gause's results to Salt's (1967) experiments with *Woodruffia metabolica*, another protozoan predator of *Paramecium*. *Woodruffia* also encysts when starved, but it does not subdivide. The result is a comparatively stable interaction.

A somewhat different aspect of the variation in vulnerability of different ages to predation is the existence of an age-size threshhold in vulnerability. The phenomenon is quite common. *Balanus cariosus*, a barnacle, is normally heavily preyed upon by the drilling snail, *Thais lamellosa*. Connell (1972) protected newly settled barnacles in the inter-tidal zone with wire cages for a number of years (Fig. 4.2). He found that *B. cariosus* adults are not eaten after they reach two years of age; *Thais* concentrates upon younger barnacles. The other major inter-tidal predator in this area, the starfish *Pisaster ochraceous*, can consume barnacles of any size class, so the invulnerability of large *B. cariosus* only extends to sites protected from *Pisaster* predation.

At first glance a threshhold of vulnerability would seem to be stabilizing. The existence of an invulnerable yet reproductively active subset of the prey population should act to minimize the likelihood of extinction of the prey. Smith and Mead (1974) found this to be the case in a simple age-structured Lotka–Volterra model. Burnett (1964) kept the granivorous mite, *Tyrophagous putrescentiae*, in the laboratory with the predatory mite, *Blattisocius dentriticus*. We see in Fig. 6.7 that this interaction tended to fluctuate but was nonetheless stable. Neither population went extinct and the amplitude of fluctuations appeared to decrease with time. Such dynamics are by no means typical of acarine predator–prey systems and suggest that stabilizing forces are at work. Burnett attributed much of the stability to the invulnerability of large prey mites. But the numbers of prey growing beyond this threshhold size would be sensitive to the relative density of predators and would, as a consequence, be expected to fluctuate over time. Such a mechanism might promote continued fluctuations rather than numerical constancy. In this case we cannot tell precisely what it would do because the predatory mite is also cannibalistic at low densities of prey. Cannibalism is a strongly stabilizing behavior and undoubtedly also contributes to the persistence of this acarine system.[3]

So far I have treated age and size variation as though they were fixed features of a species' life history. But it is only reasonable to suppose that the rates of growth and development will depend a great deal upon the quality of the environment. Murdoch (1971) examined the circumstance in which a high density of prey allows for faster growth in well-fed predators, which in turn promotes an increased rate of predation per individual predator. This he termed a 'developmental response' to prey density and suggested it should be stabilizing. Of course the reverse of this logic could also apply to the prey. More heavily harvested populations could exhibit enhanced growth rates

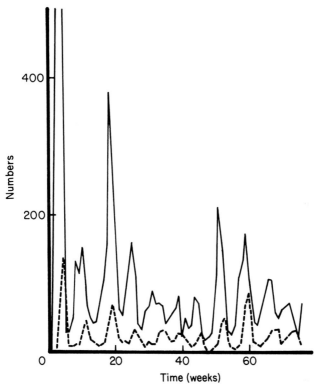

Figure 6.7   Numbers of *Tyrophagous putrescentiae* (prey – solid line) and *Blattisocius dentriticus* (predator – dashed line) in a sample of the laboratory population (from Burnett, 1964).

because of the mitigation of intraspecific competition (Kruuk, 1972). If older or larger individuals were less vulnerable to attack, then this might serve to reduce predation rates. Wilbur *et al.* (1983) may have discovered such an effect in their experimental studies of predation and competition among tadpoles. They found that predation on frog tadpoles in experimental ponds so reduced competition for food that newly metamorphosed adults were several times larger than those from predator-free ponds. The consequence is that adult frogs from heavily exploited tadpole populations are exposed to a much smaller constellation of predators between metamorphosis and sexual maturity. The implication is that increased predation on tadpoles could be directly compensated for by decreased predation on pre-reproductive adults.

To summarize, age and size structure in predator and prey populations may stabilize numbers, destabilize them, induce oscillations, or have no effect at all. Perhaps we should not be surprised that the inclusion of age-structure can produce such a range of dynamical changes. To ask what are the effects of age and size structure is to ask a poorly-posed question. Age and size structure can be manifest in a large variety of ways, some simple, some quite complicated. At present we have not treated adequately even the simple cases. The conservative conclusion is that it would be rash in any specific case to assert, without a detailed modelling effort, that this class of complications has an obvious, easily-understood effect. And to further muddy the waters, I have not treated a variety of other potentially-important kinds of variation among individuals. One would expect individual differences in physiology, disease, parasite loads, and sex to all influence both vulnerability in prey and rates of killing in predators. To the extent that such differences are heritable, they will be taken up again in Chapter 11.

NOTES

1. An elementary treatment can be found in a variety of texts, such as Mertz (1970), Krebs (1978a), Ricklefs (1979), or Elseth and Baumgardner (1981). For more complete discussions see Keyfitz (1968) or Charlesworth (1980).
2. See, for example, Orians (1969), Recher and Recher (1969), Salt and Willard (1971), Sparrowe (1972), Buckley and Buckley (1974).
3. Fox (1975), although Mertz (1969) found one strain of *Tribolium castaneum* to suffer periodic outbreaks because of variation in cannibalism.

7 Prey refugia

To human apprehension there is no balance
but a struggle in which one often exterminates
another.

Alfred Russell Wallace (1855)[1]

Our success in creating homogeneity in the human environment leads us all
too often to posit homogeneity in the natural world. But the ecological
environment almost always belies this premise. A young ecologist quickly
learns that what seems to be a monotonous stretch of marsh or forest contains
an amazing variety of habitats and that the constituent species and their
growth forms, productivities, and densities can change strikingly as one walks
within the community. In this chapter I shall explore one facet of the spatial
complexity of the natural world, that which for the prey provides relative
freedom from attack. Refuges come in three forms: those which provide
permanent spatial protection, those which provide temporary spatial
protection, and those which provide a temporal refuge in numbers. My plan in
this chapter is to examine a few empirical examples of all three types, to explore
the circumstances under which each kind is important, and to speculate upon
their implications for population dynamics.

The refuge type I shall consider first provides continuous protection for a
small subset of the prey population. Gause was the first to investigate this topic
in his work with *Paramecium* and *Didinium* (referred to in Chapter 5). He
normally used a clear oat-infusion medium for the growth of *Paramecium*, but
in one experiment he allowed a small amount of sediment to remain at the
bottom of the culture tube. Some *Paramecium* penetrated the sediment,
providing themselves with protection against *Didinium*, which remained in the
clear supernatant above. The results could hardly be said to be stable:
Didinium destroyed its prey in the supernatant, starved, and the remaining few
Paramecium in the sediment emerged to grow unchecked. But the outcome
was qualitatively different from the results without the refuge; without the
sediment, *Didinium* destroyed all the *Paramecium* before it succumbed.

Flanders conducted a number of experiments on the *Anagasta–Venturia*
system discussed in Chapter 6, in which he provided the moth larvae with a
variety of types of protection against parasitism (Flanders, 1948; Flanders and
Badgley, 1963). In one group of experiments he varied the depth of the rolled

wheat medium. Since *Venturia*'s ovipositor is about 4 mm long, an 8 mm layer of medium provides protection from attack in its lower half. In other chambers the medium was covered with a layer of vermiculite, and in still others small glass discs (12 mm diameter) were placed on top of the rolled wheat. The experiments suggested that stability is enhanced by a refuge; although an invasion of the cultures by mites and disease organisms makes the results a bit difficult to interpret. When the depth of the medium was shallow, the *Venturia* destroyed the moth population. Either increasing the depth of rolled wheat or adding glass discs mitigated this result and allowed coexistence. Vermiculite protected the larvae so well that the moth population increased to the maximum density supportable by the rolled wheat provided.

The standard explanation of this result is that by protecting a minimum number of prey, a refuge eliminates one of the two avenues by which the interaction could collapse. The logic appears to be straightforward.[2] One must modify the basic theory (Equations (2.7) and (2.8)) to make a subset S of the H prey invulnerable, i.e.

$$\frac{dH}{dt} = Hg(H) - f(P, H - S) \tag{7.1}$$

$$\frac{dP}{dt} = K[P, f(P, H - S)] \tag{7.2}$$

The phase-plane portrait of Equations (7.1) and (7.2) might appear as in Fig. 7.1. Whenever the system intersected the vertical portion of the H isocline, the resultant vector would point straight down. The predator population would keep the prey at the refuge level until P was so reduced by starvation that the prey could recolonize less-protected habitats.

Although satisfying as far as it goes, this portrait is too simple to serve as a paradigm for all refuges. The nature of a refuge and the extent of its use are not usually quite so absolute. For example, White and Huffaker (1969b) reported experiments with the same *Anagasta–Venturia* system used by Flanders. To a shallow layer of wheat flakes in Petri dishes they added either of two types of protection: small glass coverslips similar to those used by Flanders and large cards with eight punched holes. The protection provided by the cards and the coverslips was not complete; if the wheat medium subsided at high host densities, because of consumption by larvae, wasps could crawl beneath the barriers. The result was protection at low host densities and vulnerability at high host densities. The fate of these cultures was not stability but rather increased fluctuation and eventual extinction. *Anagasta* has a natural tendency to oscillate, attributable to its high fecundity and to intraspecific aggression among its larvae. The refuges apparently forced *Venturia* into the role of exacerbating these fluctuations; moth populations expanded unchecked and then were driven very low when the medium was depleted and larvae were vulnerable. White's and Huffaker's study suggests that if the

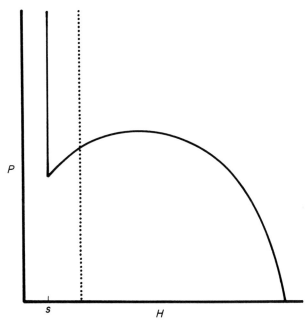

Figure 7.1   Equilibrium conditions for Equations (7.1) and (7.2) as they might appear on a phase-plane portrait. The existence of a refuge implies that the prey population will not grow smaller than *S*.

effectiveness of the refuge depends upon density, the dynamic consequences can be complicated.

The size of the refuge is also important. Had the glass coverslips been larger, *Anagasta* larvae in their centers might still have received a degree of protection. Some theoretical work on mobile populations bears on this point. By use of diffusion processes in mathematical models, a theoretician may mimic the effects of certain kinds of animal movements upon population dynamics. Since diffusion models currently comprise an important avenue of research in population ecology and are likely to become even more important in the near future, I shall sketch out here the form of the simplest model.

Imagine that the animals in a population are constrained to move in one dimension along an axis *x*. The numbers at point *x* at time *t* is $H(x, t)$. Figure 7.2 displays a hypothetical population distribution over *x* for a fixed point in time. If we assume (for the present) that animals move strictly at random, i.e. right or left with equal probability, then we can develop a portrait of how the numbers on the patch of ground stretching from *x* to $x + \Delta x$ will change over the next instant of time. The numbers in that patch are given by

$$\int_{x}^{x+\Delta x} H(s, t)\, ds \qquad (7.3)$$

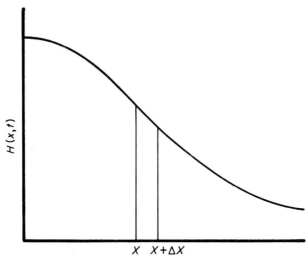

Figure 7.2   The distribution of a population over space, x, at a fixed time, t. The volume between points x and $x + \Delta x$ is the number of individuals in that space.

How this number will change depends upon the net movement of animals in or out at each boundary, which I denote with $J(x, t)$, so that the rate of change in time is

$$\frac{\partial}{\partial t} \int_x^{x+\Delta x} H(s, t)\, ds = J(x, t) - J(x + \Delta x, t) \tag{7.4}$$

The next step is to divide both sides by Δx and take the limit as Δx goes to zero. This yields

$$\frac{\partial}{\partial t} H(x, t) = -\frac{\partial}{\partial x} J(x, t) \tag{7.5}$$

The assumption that each animal moves at random implies that the movement across the boundaries should be proportional to the gradients in density at those points, or

$$J(x, t) = -D \frac{\partial}{\partial x} H(x, t) \tag{7.6}$$

Substitution of (7.6) into (7.5) gives

$$\frac{\partial}{\partial t} H(x, t) = D \frac{\partial^2}{\partial x^2} H(x, t) \tag{7.7}$$

which is the basic diffusion equation. Equation (7.7) is easily solved, and a particular solution is presented graphically in Fig. 7.3. If at some time zero all

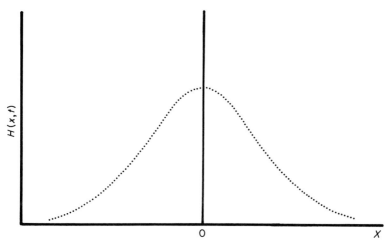

Figure 7.3   A representation of the population distribution [$H(x, t)$] as predicted by the basic diffusion model. At time zero (solid line), the population is all concentrated at $x = 0$. At a later time, t, (dotted line) the population is distributed normally with respect to x.

individuals were at point zero, then at a later time t, the population will be spread out in a normal distribution. The rate of spreading is determined by the diffusion coefficient D.

By itself this is an overly simplistic portrait of animal movement, but diffusion theory admits to a great deal more realism (McMurtrie, 1978; Nisbet and Gurney, 1982). One can generalize to two dimensions, allow the whole population to drift, provide for more complicated dispersal properties, and so forth. For the purposes of this chapter, one can examine the circumstance in which a safe habitat is surrounded by a hostile environment. An early paper in this field by Kierstead and Slobodkin (1953) came up with a striking conclusion: a population in a refuge surrounded by hostile environments would continue to grow as long as its habitat remained above a certain critical size. Once the refuge decreased below that size, the population would decline to extinction. The critical size would be determined in part by the random rate of movement of individual organisms. When a refuge is small, a sufficient fraction of its population could wander into the zone of danger that mortality would exceed reproduction. This conclusion presupposes that the animals in the refuge do not make intelligent use of it, that is they do not necessarily choose to turn back at the boundary and remain inside; however this assumption is reasonable in a variety of plausible circumstances. Gurney and Nisbet (1975) demonstrated, moreover, that this same minimum size phenomenon can occur even when the transition from a safe to a dangerous zone is gradual rather than abrupt.

Diffusion models have a variety of other applications. I shall return to them later in this chapter, but for now I shall turn to consideration of temporary refuges. These fall into two classes, those which provide real protection but which are only used during one time of the year or one life stage and those which are protected only in the sense that they remain undiscovered by predators. The first category is well illustrated by one of the better of the few field studies on this topic, Kruuk's research on the spotted hyena in east Africa (Kruuk, 1970, 1972).

Kruuk studied two population of hyenas, one in the Serengeti Plains region and the other in Ngorongoro Crater. The ecology of the Serengeti is such that during the dry season perennial grasses sequester their nutrients below ground, fires are common, and the place becomes generally inhospitable to grazers. As a consequence the great herds of wildebeest, primary prey for spotted hyenas, migrate north off the Serengeti and do not return until the dry season passes. Packs of hyenas follow these herds only to the edge of the plains. There they dig dens and wait; aversion to the tsetse fly is apparently so great that hyenas will not take their cubs into the bush. Adults will leave the plains in search of prey and return following a kill to disgorge food for the cubs. If suitable prey are not to be found close to the den site, then the intervals between meals grows too long for the cubs and they starve. On the Serengeti the dry season is a developmental bottleneck which suppresses recruitment of adults and serves to keep the predator population low.

In Ngorongoro Crater, by comparison, water is available all year round and the herds do not migrate. As a result the ratio of hyenas to prey animals is substantially greater. Kruuk estimated the annual percentage of wildebeest killed in Ngorongoro Crater to be 11.0%, by comparison to only 1.6–2.6% in the Serengeti.[3] Those wildebeest which survive this onslaught appear to be much better fed than their conspecifics in the Serengeti. We can infer from this study that a temporary spatial refuge for the bulk of a prey population can depress predator numbers and allow the prey to grow close to the limits of its food supply. Whether this would occur in other circumstances would depend upon how these circumstances would compare to the critical features of the Serengeti population of spotted hyenas: absence of prey at a vulnerable developmental stage and absence of alternative prey. Kruuk's research strategy, the comparison of the same species interaction in quite different habitats, proved to be productive and should be considered by others.

Several experimental studies point to the effects of transient spatial refuges. Without doubt Huffaker's studies of mites on oranges is the best known (Huffaker, 1958; Huffaker et al., 1963). In a variety of experimental environments, he grew populations of the six-spotted mite, Eotetranychus sexmaculatus, as prey and Typhlodromus occidentalis as its acarine predator. E. sexmaculatus feeds upon the surface of oranges, so Huffaker adjusted the complexity of his simple environments by the use of different arrangements of oranges. The numbers and spatial pattern of oranges could be varied without changing the total amount of food for the prey either by wrapping oranges in paper or by dipping them in paraffin and then exposing a fraction of their area to the mites. The first arrangements he used are described as follows:

1. Four oranges, each half exposed, in adjoining positions;
2. Eight oranges, each half exposed, grouped together;
3. Six oranges grouped together, a total of six orange surfaces exposed;
4. Four oranges, each half exposed, dispersed among 36 rubber balls;
5. Eight oranges, each half exposed, dispersed among 32 rubber balls;
6. Twenty oranges interspersed with 20 rubber balls, one-tenth of each of the oranges exposed;
7. Forty oranges, each with one-twentieth of its surface exposed;
8. One hundred and twenty oranges, with an equivalent of six orange surfaces exposed, and containing partial barriers to dispersal.

These eight treatments represent a progressive division of the food supply for six-spotted mites into smaller and more widely separated parcels. The dynamics of the mites in each of these eight experimental arenas were comparable. *Eotetranychus* increased in numbers, sometimes to very high densities. *Typhlodromus* followed with its own increase, drove the prey to extinction, or near it, and then starved to death.

In a remarkable tribute to persistence, Huffaker further complicated environment 8 by erecting wooden posts up which the prey could climb to disperse by air currents provided by an electric fan. The combination of this and different starting populations allowed the prey and predator to coexist for nearly 8 months, albeit exhibiting pronounced oscillations. His final experiment employed an arrangement of 252 oranges on three levels in a cabinet, exposing the equivalent of 12.6 orange surfaces. Mites could move from one orange to the next by means of the grids and posts supporting each level. The two species coexisted in this system for 70 weeks (Fig. 7.4). The pattern of population fluctuations was similar to that in the 120 orange dispersion.

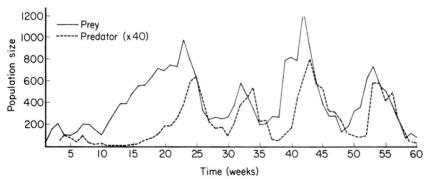

Figure 7.4 Population fluctuations of the six-spotted mite (solid line) and its predator (dashed line) on an arrangement of 252 oranges (after Huffaker *et al.*, 1963).

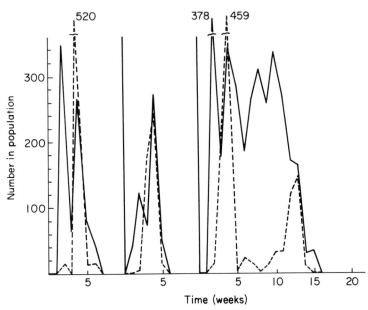

Figure 7.5 Three replicate cultures of houseflies (solid lines) and wasp parasitoids (dashed lines) in single cages (from Pimentel *et al.*, 1963).

Pimentel *et al.* (1963) followed the laboratory interaction between houseflies and their pupal parasitoid, *Nasonia vitripennis*. These were kept in various arrangements of small plastic boxes. The populations within a single box gave no signs of stabilizing (Fig. 7.5); typically the numbers of houseflies exploded, and then the *Nasonia* responded and killed them all in 5 to 15 weeks. When sixteen of these boxes were interconnected with plastic tubes, the populations persisted about 30 weeks before the wasps went extinct (Fig. 7.6). The 16-cell system was complicated further to a 30-cell system, and this allowed the populations to coexist for 83 weeks. Finally the 30-cell system was modified to discourage movement by *Nasonia*. The tubes were extended into the boxes about 2 cm and covered with petroleum jelly. Pimentel covered the ends of each tube with a screen disk containing a small opening in the center through which insects could enter. Such an arrangement did little to inhibit the movement of flies but proved to be a formidable barrier to the dispersal of *Nasonia*. The population trajectory in Fig. 7.7 suggests that this tactic produced a degree of stability. The major lessons Pimentel and his coworkers have drawn from this and subsequent related work pertain to the coevolution of hosts and parasitoids in closed ecosystems. I shall discuss this in Chapter 11 and draw attention here only to the effects of spatial structure. Given the extraordinary lengths to which both Huffaker and Pimentel were forced to find a stable laboratory system, we should not be surprised that scientists of

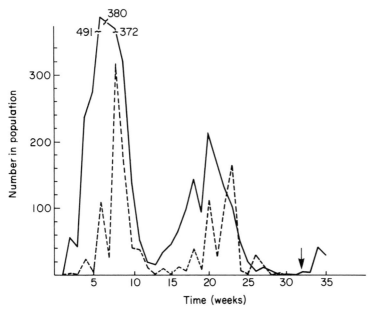

Figure 7.6 Population trajectories of houseflies (solid lines represent, here, pupae) and wasps (dashed line) in a cage consisting of 16 interconnected cells. The numbers are the mean number per cell. The wasp population died out at the time indicated by the arrow (from Pimentel *et al.*, 1963).

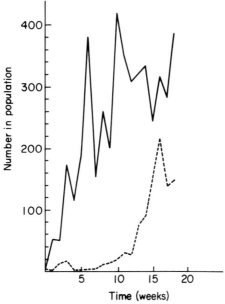

Figure 7.7 Average numbers per cell of houseflies (solid lines) and *Nasonia* (dashed line) in a 30-cell cage with barriers to parasitoid movement (from Pimentel *et al.*, 1963).

the mid-1960s often speculated as to whether or not any predator–prey systems were stable.

Takafuji (1977) performed a series of experiments with the herbivorous mite, *Tetranychus kanzawai*, and its predator, *Phytoseiulus persimilis*, cultured on potted kidney bean plants. He did not need to go to the extremes of Huffaker and Pimentel, but found nonetheless that vulnerability to extinction depended upon sufficient isolation of the subpopulations. Plants were set in pans of water in one of two arrangements (Fig. 7.8). Arrangement A featured

System A

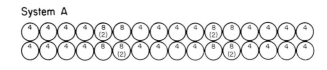

System B

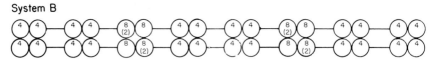

Figure 7.8   Diagram of the spatial arrangements of the kidney bean plants in the two experimental treatments. In each circle are given the number of prey mites and (in parentheses) the number of predatory mites released (after Takafuji, 1977).

32 plants located so that adjacent pots and bean plants touched. In pattern B the pots were separated into eight groups of four, and each group was connected to its adjoining groups with small wooden bridges. The prey mites dispersed little in either of these treatments, so Takafuji's experimental results are attributable primarily to movements of the predators. System B, with its much weaker linkage among the pots, persisted twice as long as system A (Fig. 7.9); both species were present in system B when the experiment was terminated on the 92nd day.

Taken in the aggregate these three studies strongly suggest that locally unstable populations, loosely coupled, can form a more stable metapopulation. The critical requisite seems to be a balancing of the dispersal abilities of the predator and the prey such that the predator neither dies out from failure to find prey nor discovers prey so quickly that they have no chance to build up locally (Hastings, 1977). In some circumstances the spatial structure required may be minimal. Maly (1978) was able to confer a degree of persistence on the notoriously unstable *Paramecium–Didinium* system by connecting together six small homogeneous populations with 1.25 mm diameter capillary tubes. Maly used *Paramecium caudatum*, not the *Paramecium aurelia* that Gause and Luckinbill had used before him, but demonstrated that in simple homogenous cultures the *P. caudatum–Didinium* system was also unstable.

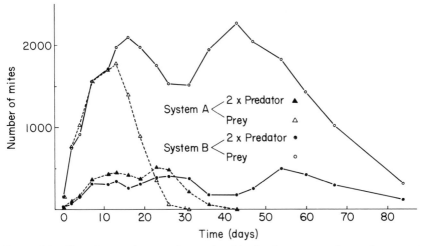

Figure 7.9 Changing predator and prey densities in the two experimental arrangements (from Takafuji, 1977).

I find it a bit difficult to interpret Maly's experiments, since he chose to run them for only a short period of time and to publish but a portion of his results. But the source of the increased persistence is clear. *Paramecium caudatum* is prone to dispersal under a range of culture conditions; whereas *Didinium* disperses well only when hungry. Well-fed individuals move little and remain near the bottom of the culture vessel. Such behavior allows *Paramecium* to establish *Didinium*-free colonies in new compartments of the experimental universe. The influence of such complicated dispersal behaviors upon the dynamics of spatially-structured populations is not well understood. In this case it looks as though it contributes to the minimum isolation necessary for stability.

All these experimental studies partitioned the environment into discrete subunits. Would the same result apply to a continuous environment of sufficiently large size? Some theory on diffusion processes suggests that spatial pattern can arise in the absence of environmental variation (McMurtrie, 1978; Hilborn, 1979). A special class of models called diffusion-reaction equations, borrowed from chemistry but applicable to predation, predicts that stable spatial structures can arise endogenously. This is an intellectually intriguing development because it suggests that the environmental heterogeneity which might be necessary for persistence of predator and prey could arise from their interaction alone. This would imply that the physical size of an environment might be sufficient by itself to stabilize locally unstable systems. Luckinbill's experiments with different sized culture vessels for the *Paramecium–Didinium* system (Chapter 5) can be reasonably explained by such a process.

The final 'refuge' I shall treat is a temporary refuge in numbers. If a prey animal cannot employ the obvious tactic of avoiding vulnerability altogether, then its next best strategy is to share the risk with its conspecifics. Such an evolutionary strategy must have led to the life-cycle synchrony found in a number of species. Synchrony of parturition in large ungulates, mast fruiting in plants, and periodicity in certain insects are different manifestations of the same tactic, the swamping of predators. The idea is quite simple: if one stage in the life cycle is vulnerable, be it young or adult, then synchronization of the development of the entire population brings everyone to that stage at once. Any population of predators has a limit to the rate at which it can kill and consume prey; those in excess of this limit will survive.

Some of these synchronized efforts provide spectacular displays. Hughes and Richard (1974) described the mass nesting of the Pacific Ridley turtle on a small beach in Costa Rica. During a 3.5 month study period, 288 000 turtles nested in three sessions of 4, 3, and 3 days. Only 2178 nested in between. A large fraction of the approximately 11.5 million eggs laid during one of these mass nestings apparently never hatched. The hatchlings from the remainder had to run a gauntlet of black vultures, frigate birds, and ghost crabs. Only about 0.2% of the eggs resulted in a hatchling reaching the water, but this is probably a much greater rate of success than would occur if the process were not synchronized.

Such patterns are common in ungulates. Kruuk observed that births of wildebeest calves all occur in a three-week period from mid-January to mid-February. Murie (1944) found the same pattern in Dall sheep in Alaska. Neither hyenas in the first case nor wolves in the second were able to kill all the young before they were sufficiently mature to run well.

The precise synchrony of North American periodical cicada populations is best explained as a response to predation. Variants appearing earlier or later than their 13- or 17-year-old conspecifics would find a high ratio of predators to prey and would be at a selective disadvantage (Lloyd and Dybas, 1966).

The key to the successful operation of this kind of temporal refuge is the inability of predator populations to build in anticipation of the availability of prey. Large carnivores eat well during the calving season, but their numbers are constrained by the availability of food at other times of the year. Predators of cicadas are short-lived and would need to build up numbers on alternate prey. Even were this possible, the timing mechanism would be awkward since the periods between cicada emergences are prime numbers. To prepare for an outbreak of 17-year cicadas, a predatory species must have either an annual or a 17-year generation time; it is difficult to conceive of a periodic burst in reproduction carrying over 17 years.

The construction of evolutionary scenarios for the phenomenon of synchrony of vulnerable periods comes substantially easier than does an understanding of the dynamic consequences. The sheer intensity of killing during these periods makes it unlikely that any sort of subtle regulatory adjustments in predation rates are occurring. Kruuk observed a pack of hyenas stumble by chance upon a calving ground of Thompson's gazelles.

What followed was simply mass slaughter of the newborn antelope, with no consideration of the nutritional needs of the hyenas.

During cicada outbreaks, birds and mammals engorge upon the large and energetically valuable adult insects. But their actual numerical effect, in spite of the carnage, is probably trivial. Cicada numbers are more likely limited both by competition for space and food among the nymphs, which live underground, and by parasitism by the specialized fungus, *Massospora cicadina* (Soper *et al.*, 1976; White *et al.*, 1979). The consequences of this synchrony for long-term dynamics are too poorly understood to allow us to advance much in the way of generalizations. One can only observe that by this mechanism some prey species are able to coexist with their predators even though they have completely helpless stages in their lives.

NOTES

1. Quoted in McKinney (1966).
2. Rosenzweig and MacArthur (1963), St. Amant (1971; cited in Murdoch and Oaten, 1975).
3. Lions, the other major wildebeest predators, occurred in both locations in the same ratios of predators to prey.

8 The functional response: the influence of predatory behavior upon dynamics

The human mind is incapable of thinking
other than about models.

Isaacs (1979)

Several decades ago, Solomon (1949) chose to approach the question of predation's role in population dynamics by partitioning the response of predators to changing prey numbers into a numerical change in predator density and a functional change in the rate of predation per individual predator. In other words, predators can respond to an increase in numbers of prey both by increasing their own numbers and by eating more per predator. In the context of the basic theory used in this book, Equations (2.7) and (2.8), the numerical response is the Equation (2.8), and the functional response is the function $f(H)$. This chapter will discuss $f(H)$, the factors that determine its shape, and its implications for population dynamics.

Solomon's dichotomy has not gone unchallenged. Hassell (1966) suggested an alternative division of the predatory response between behavioral and inter-generation responses. He argued that some numerical responses by predators are long-term, such as the alteration of rates of survival and reproduction, and others are short-term, such as behavioral aggregation in areas of high prey density. The original terms were sufficiently well-established by that point that Hassell's alternatives failed to displace them, but his point is well taken. The time lag in the predatory response is important and should be acknowledged. And, partly in response to Hassell's suggestion, most ecologists today gather all short-term behavioral responses under the rubric of the functional response.

Solomon's dichotomy proved productive primarily in its inspiration of a series of influential papers by C. S. Holling, in which he explored the causes and implications of variously-shaped functional response curves.[1] Figure 8.1 contains the four broad classes of functions Holling termed types I–IV. Common to all four is an upper bound on the rate of predation per predator at some prey density, in contrast to the old Lotka–Volterra model which had

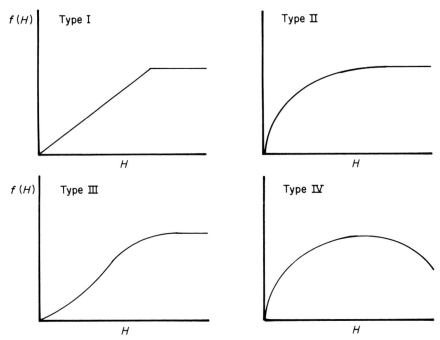

Figure 8.1   Holling's four classes of functional response curves.

assumed a linear relationship between prey density and the rate of predation over the entire range of prey densities. Several theoreticians recognized linearity in $f(H)$ as unreasonable at the outset, Volterra among them (Volterra, 1928); however, the need for an upper limit on the rate of predation was not commonly recognized in the 1950s. The curves of types I, II and III reflect different patterns of monotonic rise to this bound; the type IV curve exhibits, in addition, a decrease in the predation rate at high prey densities. Experimentalists have discovered all four types of curves in laboratory systems; in fact, the description of functional response curves for this or that experimental animal became something of a cottage industry in the 1970s (reviewed by Hassell, 1978).

The consequences for population dynamics of the shapes of these curves have been thoroughly explored over the last couple of decades by means of the insertion of a variety of mathematical models of the functional response into various sets of predator–prey equations. The conclusions have all been comparable and qualitatively similar to Holling's original heuristic analysis. I shall recapitulate his logic first and then present one example of a more precise analysis.

Holling compared the four types of curves by the simple criterion of whether or not the predator population would meet an increase in prey numbers with an increase in the percentage of the prey population eaten per predator per

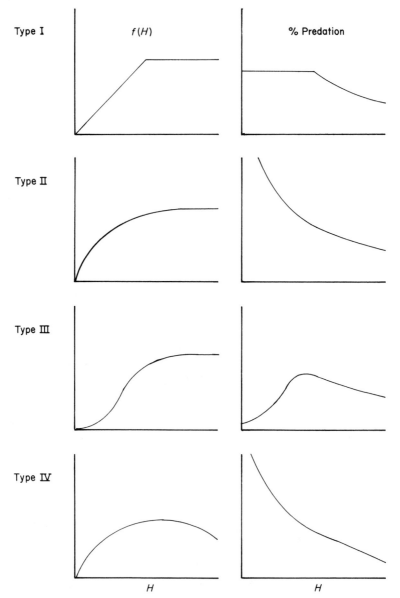

Figure 8.2 The percentage of a prey population killed by an individual predator as a function of prey density for each of the four classes of functional response curves.

unit time. Such an increase in the percentage eaten would tend to be stabilizing, although it might not be enough to actually produce stability. In the curves of types I, II and IV, the percentage either remains constant or declines with increasing prey numbers (Fig. 8.2). Only the sigmoid curve, type III, has stabilizing potential and this only if an equilibrium exists within the region where it is accelerating.

Gause *et al.* (1936) and Murdoch and Oaten (1975) approached the question more precisely by embedding the function, $f(H)$, in the Lotka–Volterra equations, to give

$$\frac{dH}{dt} = aH - Pf(H) \tag{8.1}$$

$$\frac{dP}{dt} = -bP + cPf(H) \tag{8.2}$$

The results of their stability analysis can be portrayed graphically (Fig. 8.3). One locates the equilibrium abundance of prey on the abscissa, draws the tangent to $f(H)$ at that point, and determines the point of intersection of the tangent on the ordinate. The criterion for stability requires that the intersection be below the origin, a criterion obviously met only by type III curves. This graphical approach reveals more precisely than Holling's discussion the region of prey densities in which type III curves are effective, but its precise predictions must depend upon the choice of equations in which to embed $f(H)$. Oaten and Murdoch (1977) defend their use of the Lotka–Volterra equations as a vehicle for the exploration of the shape of the functional response, with the argument that their goal was not so much prediction of the precise effect of a particular functional response curve, but rather the exploration of broad patterns accompanying changes in the shape of the curve. By implication, those general qualitative changes should occur in the same direction regardless of the set of equations chosen to hold $f(H)$. This is a worthwhile claim to ponder, but it ignores the equally valid claim that we need to understand how reasonable and general classes of functional response curves interact to influence stability with reasonable forms of the other remaining functions in a general model. Levin (1977), for example, has shown that proper choice of the self-limiting function, $g(H)$, in Equations (2.7) and (2.8) can produce stability in the face of a destabilizing functional response.

Armstrong (1976) addressed this issue by offering a more general version of the same sort of graphical analysis. He chose to insert a general functional response in the following set of equations:

$$\frac{dH}{dt} = G(H) - Pf(H) \tag{8.3}$$

$$\frac{dP}{dt} = PK(H) \tag{8.4}$$

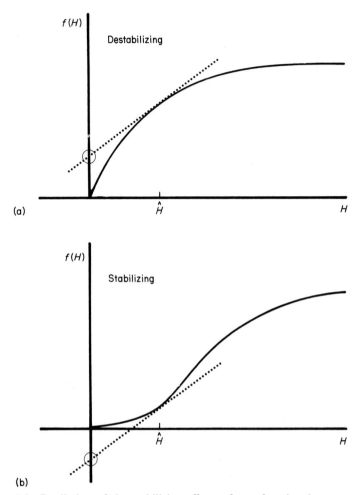

Figure 8.3 Prediction of the stabilizing effects of two functional response curves evaluated according to the method of Gause *et al.* (1936) and Murdoch and Oaten (1975). The tangent to the curve (a) at the equilibrium intersects the ordinate above the origin, implying destabilizing effects. The tangent to curve (b) intersects below the origin and will tend to produce stability.

These differ from Equations (2.7) and (2.8) in that the form of the prey growth rate in the absence of predators, $G(H)$, allows for more possibilities than does the term $Hg(H)$, and the removal of P from inside the parentheses in the second term of the first equation, and in the second equation, eliminates the possibility of predators interfering with one another, other than indirectly by influencing the food supply. Without question, however, they are more general than the Lotka–Volterra equations. Equations (8.3) and (8.4) are identical to those used by Rosenzweig (1969) in his isocline analysis; Armstrong's procedure is simply another way of looking at the same model. It has the advantage that $G(H)$ and $f(H)$ can be separated. The procedure requires first that the functional response $f(H)$ be plotted on the same axes as the growth rate of the prey $G(H)$ (Fig. 8.4). Then the equilibrium prey density, H, is located on the abscissa, and tangents to the curves at that point are drawn to their intersection with the abscissa. The stability of the interaction can be inferred directly from the two H-intercepts, I_F and I_G, meeting the condition

$$\frac{1}{\hat{H}-I_G} < \frac{1}{\hat{H}-I_F} \tag{8.5}$$

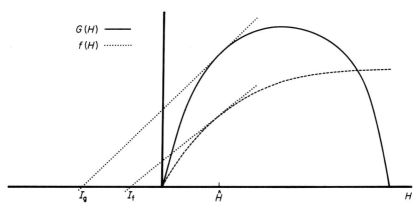

Figure 8.4 Armstrong's graphical procedure for evaluating the stability of a general set of predator–prey equations. $G(H)$ is the population growth rate of the prey in the absence of predators; $f(H)$ is the functional response. Conditions for stability emerge from the relative values of I_g and I_f, the intersections with the abscissa of tangent lines to these curves at the equilibrium, $\hat{H}$ (see text).

From Inequality (8.5) we can infer the following:

1. if the prey growth rate at equilibrium is hampered by crowding ($I_G > \hat{H}$), then the interaction will be stable for any type I, II or III curve;
2. if the prey are sparse, such that $I_G < \hat{H}$, then stability will only result if $I_F > I_G$.

Note that the second condition does not require that the functional response necessarily be sigmoid, although that would certainly ease things. Figure 8.5 shows two different functional responses analyzed in this fashion, one resulting in a stable system and one not.

Before proceeding from an analysis of stability to discussion of the factors influencing the shape of the functional response, I must emphasize that stability and sigmoid functional responses do not always march hand-in-hand.

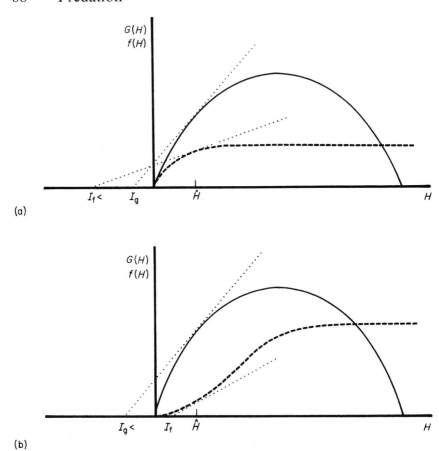

Figure 8.5 Demonstration of how, for a given $G(H)$ function and a low $\hat{H}$, one functional response curve (a) will result in instability; whereas another (b) will produce stability (see text for explanation).

You will find stability or instability in systems displaying type III functional responses, and a type I or II functional response can be found in a perfectly stable interaction. Stability or the lack of it is an emergent property which cannot be inferred from one aspect of predation taken alone.

How do these curves arise? The types I and II curves are the easiest to explain. Type I was originally conceived as a simple modification of the linear functional response of the Lotka–Volterra equations. The model predator generating such a curve would be a filter-feeding zooplankter harvesting single-celled algae at a constant rate until its gut was full. The type II curve reflects a more general saturation process and is typically generated in either of two ways. In the first model the rate of feeding is assumed to decline continuously as a function of the filling of the gut, going to zero as the gut becomes full. This explanation, advocated independently by Gause (1934) and Ivlev (1961), derives primarily from consideration of the effects of hunger upon the motivation to hunt. The model they chose to describe this phenomenon assumes a linear decrease in feeding rate as it approaches a maximum:

$$\frac{df(H)}{dH} = a[f_{max} - f(H)] \tag{8.6}$$

The two constants represent the rate of predation when hungry (a) and the maximum rate of predation (f_{max}). It is not clear why satiation should produce such an alteration of feeding rates, but the simplicity of Equation (8.6) is probably sufficient justification for its use. When integrated, Equation (8.6) yields

$$f(H) = f_{max}(1 - e^{-aH}) \tag{8.7}$$

The second model builds upon the observation that the handling and digestion of prey are time-consuming processes. The upper bound on the predation rate is presumed to result from a limit on the total amount of time available for feeding activities. This is Holling's original explanation for the type II functional response. His inspiration for the handling-time mechanism was a series of experiments in which a blindfolded subject picked up paper discs from a table; consequently his model is known as the 'disc equation' (Holling, 1959b).

Holling partitioned the time available for hunting into that spent searching, t_s, and that spent capturing, handling, and digesting prey once found, t_h. If prey are discovered at a rate a when searching, then in t_s seconds, the number of discovered prey will be aHt_s. Since the actual predation rate is the number of discovered prey divided by the total time for search and handling, it equals

$$f(H) = \frac{aHt_s}{t_s + t_h} \tag{8.8}$$

But if h is the average time spent handling each prey, then $t_h = aHt_s h$. Substituting this into Equation (8.8) for t_s and manipulating yields the disc equation

$$f(H) = \frac{aH}{1 + ahH} \tag{8.9}$$

which is identical in form to the Michaelis–Menten model of enzyme kinetics and the Monod formula for bacterial growth.

Some debate has occurred over which is the proper approach to modeling saturation of $f(H)$, the Gause–Ivlev or the Holling model. Both models generate type II curves, and I must emphasize that the stability of the interaction depends only upon the shape of the curve, not the process that produced it. For the sake of those who are staunch advocates of one or the other, I shall rederive the disc equation in such a way as to help reconcile the two approaches and also identify more clearly the assumptions underlying each.

Conceive predation as a queuing process; a queuing system is simply a line of customers waiting for service at a counter. Suppose that those customers are prey and the service consists of being eaten and digested. The predator is the server behind the counter. Its possible states are only two: it may either be occupied with the handling or digestion of prey (denoted by 'f' for full) or it may be searching (s). Assume that prey are encountered randomly at rate λ when no customers are being served (the predator is searching) but that prey are impatient customers and do not wait when the predator is busy. Assume also that prey are processed, once caught, with mean rate μ. In a very short time interval, Δt, only three possibilities exist, a full predator completes handling the prey and begins to search, a searching predator makes a capture, and whichever state the predator is in remains unchanged. The probability of searching or being full at time $t + \Delta t$ can be captured mathematically with the following equations:

$$P_s(t + \Delta t) = P_s(t)[1 - \lambda \Delta t] + P_f(t)\mu \Delta t \qquad (8.10)$$

$$P_f(t + \Delta t) = P_s(t)\lambda \Delta t + P_f(t)[1 - \mu \Delta t] \qquad (8.11)$$

And I shall also assume that

$$P_s(t) + P_f(t) = 1 \quad t \geq 0 \qquad (8.12)$$

By suitable manipulation of Equations (8.10) and (8.11), one can get differential-difference equations of the following form:

$$\frac{dP_s(t)}{dt} = -P_s(t)\lambda + P_f(t)\mu \qquad (8.13)$$

$$\frac{dP_f(t)}{dt} = P_s(t)\lambda - P_f(t)\mu \qquad (8.14)$$

These equations are linear and soluble, therefore one can derive probability densities for $P_s(t)$ and $P_f(t)$. Here I shall only examine the equilibrium solution of Equations (8.13) and (8.14). By setting the derivatives equal to zero, one gets $P_f = P_s(\lambda/\mu)$ for both equations. By substitution into Equation (8.12),

$$P_s = \frac{1}{1 + (\lambda/\mu)} \qquad (8.15)$$

Since prey can be caught only when the predator is searching, it must follow that the expected rate of killing is λP_s, or

$$f(H) = \frac{\lambda}{1 + (\lambda/\mu)} \tag{8.16}$$

which is the disc equation for $\lambda = aH$ and $\mu = 1/h$.

The point of working through this more elaborate derivation is to get at some of the assumptions underlying this process. Sjöberg (1980) has shown that the disc equation emerges as a special case of a more general treatment of predators as servers in a queuing system. By a heuristic examination of Sjöberg's model we can more easily see how the various curves of types I and II emerge from different biological circumstances. To follow his logic, we modify the simple formulation to allow prey, once caught and swallowed, to enter a digestion queue of maximum length R. The parameter R is a measure of gut capacity and reflects the ratio of the size of the predator to the size of the individual prey. The prey proceed through the queue to digestion, which service is performed on the prey one at a time at a fixed rate per volume. Figure 8.6 displays a class of functional response curves, in which both the rate of feeding and the density of prey are normalized to the maximum gut volume R. This set of curves reflects the assumption that the feeding rate is either constant when the gut is not full, or zero when it is. When the gut capacity is one, the model reduces to the simple case already discussed, the disc equation. As the gut capacity increases, the functional response converges to the stepwise type I curve. Such a curve fits nicely the standard model of a filter-feeding zooplankter consuming very small food particles. The point to be gained from Fig. 8.6 is that these two curves represent the extremes of an infinite variety of curves spanning the range ($1 \leq R < \infty$).

Sjöberg explored a second set of models in which the rate of feeding, λ, is no longer considered a constant but declines according to the formula

$$\lambda_i = \frac{\lambda_{max}}{i + 1} \tag{8.17}$$

in which i is the number of food items in the gut. This was intended to mimic the effects of satiation lowering the motivation to hunt. In the special case where the gut is assumed to be limitless ($R \to \infty$), one gets the Gause–Ivlev model (Equation (8.6)). The functional response quickly converges to this form for surprisingly small values of R, as Fig. 8.7 shows. Figure 8.7 also shows that the Holling and Gause–Ivlev models are identical when the gut capacity is one.

When, as in this circumstance, one can derive various specific models as special cases of a more general formulation, it is not so profitable to argue about which specific model is best as it is to discover the assumptions which lead to one or the other. I re-emphasize that since all these curves will be of type II and since their effects upon stability have been shown to depend upon shape

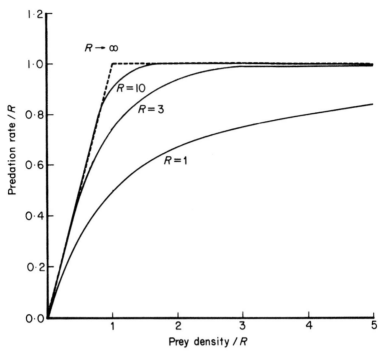

Figure 8.6 Scaled functional response curves for queuing models in which gut capacity, R, ranges in value from 1 to infinity. The shape of the curve for $R = 1$ is that predicted by Holling's disc equation (after Sjöberg, 1980).

alone, their differences in functional form probably matter little to population dynamics.

The search for the biological roots of type III curves has been substantially more confusing. Around 1960, Holling and L. Tinbergen both published papers on predators which displayed type III functional responses (Holling, 1959a; Tinbergen, 1960). Holling's study described depredation of cocoons of the European pine sawfly in both field and laboratory studies in Canada; Tinbergen investigated titmice in the Netherlands, whose prey were lepidopteran larvae in a pine plantation. Both the mammalian and avian predators showed a lag in the time between the appearance of a prey species in the environment and its subsequent appearance in the diet, and both investigators invoked learning to explain this pattern. Tinbergen coined the now-famous term, *specific search image*, to characterize how his birds learned searching skills. The idea of a specific search image, a learned visual filter, had a great deal of appeal to animal behaviorists. Curio (1976) gives a good review of subsequent research on the topic. Although the idea at the time had an aura of novelty, the process is simply one of operant discrimination learning (Shettleworth, 1972).

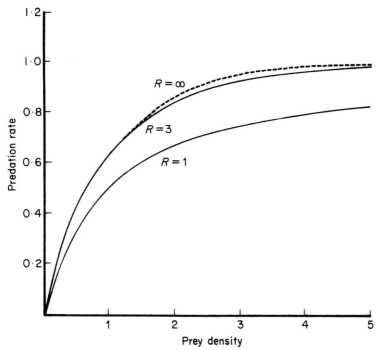

Figure 8.7 Functional response curves generated by queuing models when the rate of feeding depends upon the number of prey in the gut. The Holling disc equation results when $R = 1$; the Gause–Ivlev model when $R \rightarrow \infty$ (after Sjöberg, 1980).

It is worth noting that neither Holling nor Tinbergen actually observed learning directly; they documented a transition from one prey species to another, such that the more common type represented a larger part of the diet than would have been expected. From this phenomenon, now called switching (after Murdoch, 1969), they inferred that learning must have occurred. Hindsight and the subsequent development of switching research in the 1970s suggest that both Tinbergen and Holling may have been wrong.

It should come as no surprise that learning can indeed produce switching, but learning can occur at several stages of the predation process, not only the search for individual prey. For example, the location of prey might involve the use of a hierarchy of initially neutral habitat cues, in addition to the cues emanating directly from the prey individual, any or all of which a predator could learn to associate with the presence of prey. Stalking, chasing and subduing prey are skills which may require a great deal of experience to perform well. Even the digestion of prey may be initially slower until the appropriate enzymes are produced in sufficient quantity.

That all of these steps will be important in the generation of a type III response in any specific case is unlikely; many have insufficient latitude for improvement. I have suggested elsewhere (Taylor, 1974) that searching cues be ranked according to their value, value being defined as the difference in the rate of predation before and after a specific cue or skill is learned. For example, suppose that titmice hunt a geometrid caterpillar which is found on new leaves of a particular species of oak. The larvae are cryptically colored as well. Three skills need to be learned: to hunt on oaks of the correct type, to search new leaves, and to pick out the larvae from their background. The three skills interact as the product of experience-dependent probabilities to determine the actual probability of capture, i.e.

$$P(capture) = P(hunting\ on\ oaks) \times$$

$$P(searching\ new\ leaves/hunting\ on\ oaks) \times$$

$$P(discovering\ a\ larva/encountering\ it) \times$$

$$P(encountering\ a\ larva/searching\ new\ leaves)$$

Any of the first three factors will improve with experience from a lower level to a higher. The difference between the two levels is what is meant by value. Suppose the probabilities assort as in Table 8.1. In all three cases learning has

Table 8.1 Hypothetical variation in searching success as a function of experience (see text).

Skill	Probability of success		
	Inexperienced	Experienced	Value
1. Hunting in oaks	0.02	0.85	0.83
2. Searching new leaves	0.70	0.90	0.20
3. Discovering a larva	0.35	0.50	0.15

occurred, but the most valuable skill to acquire is clearly searching on oaks. I suggest the most valuable skills reveal those most likely to generate a type III functional response.

The type III response need not be restricted to switching predators. A parasitic wasp may attack only one host species in its lifetime; yet it may need to learn some of the habitat cues necessary to discover that host effectively. When one considers that many parasitoids have a short generation time, in some cases shorter than their hosts, then such a learning period could produce a sigmoid functional response (Taylor, 1974; Hassel et al., 1977).

Royama (1970) launched the first serious criticism of the learning model with his research on titmice in England. He also observed switching in the diets of these birds but accounted for it not by the formation of specific search images, but by changes in the locations in which the birds foraged. He concluded, correctly, that the observation of switching does not by itself constitute evidence for the formation of a specific search image, that alterations in the use of foraging locations differing in prey species composition could also produce switching.

I suspect that Tinbergen recognized this possibility himself, at least he observed that 'in any case, one other obvious explanation, namely that at first the birds were not hunting in the places where *Acantholyda* lives, can be discounted. In both years [in which switching was observed] they were feeding in the crowns from the start' (Tinbergen, 1960, p. 309).

But Royama's major contribution lies not in his indictment of Tinbergen's perhaps facile invocation of search image formation as the source of the switch; it lies rather in his observation that alteration of patch use, and the switching it might imply, could be the result of the rational choice of the best foraging site and not be a learning process at all. Learning as a process of information gathering might not be nearly as important as Holling and Tinbergen thought. The idea that choice of foraging site might be an optimization process was timely because it coincided with the first efforts by community ecologists to explain niche dimensions with optimal foraging arguments (MacArthur and Pianka, 1966; Emlen, 1966). The use of optimization arguments in the intervening decade has led to an improved understanding of the behavioral roots of switching. Much of the older work has been re-interpreted and is now better understood.

As an example of this, I offer Holling's original laboratory experiments on deer mice preying upon buried cocoons of the European pine sawfly. In his original report (1959a) he observed that when offered a range of densities of cocoons and a constant quantity of dog biscuits as alternative food, the captive deer mice exhibited a sigmoid functional response (Fig. 8.8). His experiments employed three mice, each of which was first allowed to hunt for sawfly cocoons in the arena for one week before experiments began and then presented with one density of cocoons for 24 hours, for a total of about 40 experimental replicates per mouse.

In a subsequent paper (1965) Holling chose to attribute the results of these experiments to learning. He argued that at high densities of cocoons the mice were successful so often that the decay in the link between the stimulus of the odor of a cocoon and the response of digging it out of the sand was trivial. When prey were sparse, reinforcement of this stimulus–response link was not nearly so frequent, and the mice forgot how to hunt. For anyone who has watched the learning behavior of small mammals, this is an implausible explanation. Mice learn rapidly, to be sure, but they do not forget nearly as fast. A mouse which has been hunting cocoons in an arena for 30 or 40 days is unlikely to forget how to find them on the day of a low experimental density.

A more likely explanation is that the mice knew perfectly well that sawfly cocoons were buried in the sand after only one or two discoveries. They simply

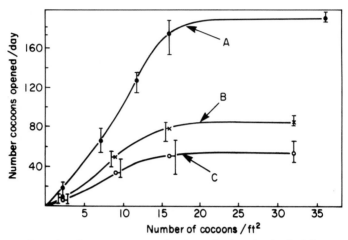

Figure 8.8 Functional response curves for deer mice hunting in the laboratory. Each curve represents a number of trials with a single mouse. The error bars show the range in performance at a given density (from Holling, 1959a).

chose not to dig them out at low densities, because the activity was unprofitable. As Abrams (1982) has pointed out, foraging has costs as well as benefits and in unrewarding circumstances the more profitable tactic may be to simply stop hunting. In experiments with grasshopper mice hunting buried prey, I observed animals abandon foraging activities on a number of occasions when the rate of return became too low (Taylor, 1977). If deer mice with low densities of prey choose to forage for 2 hours out of 24, and deer mice with high densities of prey choose to hunt for 8 hours, then the computation of a rate of predation per day will inevitably result in a type III functional response. Learning may indeed have been involved, but only in the brief assessment of what number of prey the experimenter provided that day.

In the enthusiastic pursuit of a new idea, scientists are often tempted to discard an earlier explanation, perhaps too hastily. Such seems to have happened in the last decade on this topic; behavioral ecologists searching for the roots of type III responses have embraced optimal foraging arguments wholeheartedly and have relegated learning to a minor role. Too few scientists recognize that the issue in the comparison of these approaches to predatory behavior is not whether learning or optimal foraging is important. To consider these as alternatives to each other is akin to considering the driving of an automobile an alternative to the reading of a road map; the one activity is mechanistic, the other strategic. One could legitimately argue about the relative roles of information gathering and information use in prey switching but not about learning versus optimal foraging.

Murdoch and Oaten (1975) suggest that switching can also result from the expression of strong but transient preferences. Whether their assertion adds much to our understanding of the roots of this process depends almost entirely upon how one chooses to define preference. The term has been bandied about quite loosely since its first use in ecology. Most traditional definitions are phenomenological, preference being the result of this or that experiment. Some have defined it as a choice from a menu of prey items presented simultaneously, displayed perhaps as a ranking of the set. Others use a maze of one sort or another, in which the arm of the maze chosen by the predator is claimed to reflect a preference for the prey contained therein. A third set of researchers employs specific attack behaviors. For example, Holling (1964) considered preference by mantids for flies of different sizes to be revealed by the proportion of presentations eliciting attacks of a range of sizes of models. Beukema (1968) described preference in stickleback fish as the fraction of a prey type which, upon capture, was ingested rather than spit out.

Definitions such as these have drawbacks. They emphasize choice at a small subset of the behavioral phases of the predation process and are usually meaningful only in the context of a particular experiment. Many require either model prey or various unrealistic manipulations of real prey to make the experiments tractable. For these reasons and because field biologists rarely have access to these kinds of data anyway, most ecological researchers use what I choose to describe as a black-box definition of preference. This is a quantitative description of a deviation of the diet from what would have been expected on the basis of the availability of different prey types in the environment. Suppose, for example, that someone conducting research on owls collects pellets for several weeks and discovers 25% of the remains to consist of voles and 75% of deer mice. But simultaneous trapping in the same area yields 75% voles and 25% deer mice. The standard interpretation of such data is that owls prefer deer mice.

But such an interpretation of preference merely amounts to a quantification of our ignorance of what actually occurs in prey choice. By this definition, preference reflects every decision made by a predator: to search, to capture and to consume. It must reflect any disparity between our definition of the predator's environment and its own choice of habitats, between our identification of the constellation of suitable prey types and the predator's, and between the set of prey we perceive to be discoverable and those actually within the predator's perceptual limits. If we have information on habitat use and searching behaviors, we might reasonably narrow down the factors contributing to prey preference, but the black-box definition must retain all residual factors contributing to the items actually included in the diet. In a sense, therefore, to label a switch in diet as the expression of a transient preference is merely to affirm one's chosen definition of preference. The behavioral roots of the switch remain no less obscure.

Such black-box definitions of preference have served us well in one respect. Because they have led to quantitative indices, they enable us to establish criteria by which switching may be detected. For this, one needs to establish at the minimum the null hypothesis of a fixed preference from which a switch is to depart. A variety of good black-box indices exist, and a variety of poor ones as well.

Of those in the second category, the most widely used has been the electivity index of Ivlev (1961). The proportion of a prey type in the diet, r, is compared to its proportional representation in the environment, p, according to the following formula

$$E = \frac{r - p}{r + p} \qquad (8.18)$$

Electivity ranges from -1 to $+1$ with a value of zero taken to mean no preference. Its fundamental flaw is that no data so evaluated can ever demonstrate a strong electivity either for or against a prey species when both r and p are high. This point has been made by Jacobs (1974) and several subsequent workers,[4] but electivity continues to be used.[5]

Lechowicz (1982) has reviewed the various acceptable indices. My preference is the 'alpha-vector' proposed by Chesson (1978). This is the only measure which avoids the artificiality of either limiting choice to only two species of prey or lumping all but one type into the category of 'alternative prey'. Chesson defined, for m species of prey, a vector

$$\boldsymbol{\alpha} = \begin{bmatrix} \alpha_1 \\ \alpha_2 \\ \vdots \\ \alpha_m \end{bmatrix} \qquad (8.19)$$

the individual elements of which measure the deviation of the probability of eating prey type i from the expected value of p. The α_i are normalized, such that $\Sigma \alpha_i = 1$, to comprise a set of relative preferences for the m species. Chesson provides a probability density function for the total number eaten and an intuitively reasonable maximum likelihood estimator for the α's (Chesson, 1983).

The point of worrying about quantitative measures is that the detection of switching under natural conditions can be quite difficult. Experimental studies in the laboratory have demonstrated switching by offering an extreme range of relative prey densities (Murdoch, 1969; Cornell and Pimentel, 1978), a manipulation usually impossible for the field biologist. But given that we can estimate α-vectors before and after a shift in the prey environment, it seems reasonable that for a given density and diet we should be able to reject statistically the null hypothesis that no change in preference, i.e. no switch, has occurred.[6]

To summarize, type III functional responses can arise from the learning of new skills, from optimal shifts in the use of existing skills, or from a variety of other things which, out of ignorance, we call expressions of preference. Since most such processes will accompany a change in diet from one type of prey to another, they may not translate directly into numerical changes in the abundance of predators. As a consequence the stabilizing influence of a type III response will usually occur only in the vicinity of the prey's equilibrium density. Once the prey have grown to much higher densities, predators are swamped and any stabilizing effects resulting from alteration of predatory behavior are lost.

To re-emphasize a point I made earlier, little justification exists for assuming *a priori* that all predator–prey systems of interest are stable. And if a particular system is stable, no justification exists, whatsoever, for assuming that the functional response is necessarily sigmoid. The ultimate stability properties of a system depend upon all aspects of the life histories of the component species and their interaction, not just upon predatory behavior.

Conspicuously absent from this discussion of the biological roots of functional response curves is the dome-shaped or type IV response. Its strongly destabilizing effect has apparently made it an esthetically unattractive alternative. In the next chapter I examine several circumstances in which it might arise.

NOTES

1. Holling (1959a, 1959b, 1961, 1963, 1964, 1965, 1966).
2. This line of reasoning is a simplification of a derivation in Curry and DeMichele (1977).
3. To be rigorous, Equations (8.10) and (8.11) should be written

$$P_s(t+\Delta t) = P_s(t)[1 - \lambda\Delta t] + P_f(t)\mu\Delta t + o(\Delta t) \qquad (8.20)$$

$$P_f(t+\Delta t) = P_s(t)\lambda\Delta t + P_f(t)[1 - \mu\Delta t] + o(\Delta t) \qquad (8.21)$$

The $o(\Delta t)$ represent additional terms of order Δt which will go to zero quickly as Δt becomes small. These equations are rearranged to

$$\frac{P_s(t+\Delta t) - P_s(t)}{\Delta t} = -P_s(t)\lambda + P_f(t)\mu + o(\Delta t)/\Delta t \qquad (8.22)$$

$$\frac{P_f(t+\Delta t) - P_f(t)}{\Delta t} = P_s(t)\lambda - P_f(t)\mu + o(\Delta t)/\Delta t \qquad (8.23)$$

the limits of which, as $\Delta t \to 0$, approach Equations (8.13) and (8.14).
4. Paloheimo (1979), Strauss (1979), and Lechowicz (1982). I find this problem most easy to visualize if I transform a data point on the r–p plane to a new coordinate system (r^*, p^*) rotated 45° counterclockwise, so that the p^* axis coincides with the original no-preference line. In this new coordinate system, Ivlev's index is the ratio of r^* to r^*_{max}, the maximum

possible deviation from the original no-preference line. By constructing such a graph, the reader can easily show that electivity is a poor measure of preference for all data such that $1 \leq r + p \leq 2$.

5. Zaret (1980), Pastorok (1981), Mann (1982).

6. Various procedures are available (Lechowicz, 1982).

9 Spatial structure in prey populations

There can be no room for doubt that it is
suitable to the well-being of cattle in a country
infested with beasts of prey to live in close
companionship.

Galton (1883)

Chapter 7 dealt in part with a class of predator–prey systems in which spatial structure provided a degree of stability, as a consequence of relatively limited movements by a voracious predator. The dispersal of prey from untouched populations to empty habitats counteracted the unhesitating devastation by a predator of its local prey base. But this result must apply only to those spatially-structured systems in which the dispersal abilities of the protagonist species are delicately balanced. Too mobile a predator and the prey lose their necessary sanctuaries; too mobile a prey and the spatial structure evaporates as the predator loses any reason for abandoning an area.

In this chapter, I treat the circumstance in which prey aggregate on a fine scale relative to the predator, so that the predator's hunting environment is not spatially homogeneous. The most conspicuous and dramatic examples of aggregation are of this form, great herds of ungulates, miles-long schools of herring, seabirds nesting by the thousands. And I address here the following two questions: (1) to what extent do prey choose to aggregate in response to predation, and (2) what are the dynamical consequences of this choice?

The utility of the premise that organisms are randomly spaced lies primarily in its service as a null hypothesis for comparison to their real dispersion. Reality typically corresponds to aggregation. One reason for this is that almost no species finds all of its environment equally suitable. Another reason is that animals may need to aggregate periodically for purposes of mating or nesting. Such behaviors may be sufficiently necessary for successful reproduction that they will occur in one form or another regardless of the consequences for predation. Other patterns of aggregation will result from the fact that all animals are, in one sense or another, predators themselves. When their food is clumped they must actively or passively gather together to eat it. The processes of searching are sometimes facilitated by grouping together. Krebs *et al.*

(1972) discovered that titmice searching in groups of four were more likely to find food than birds hunting alone or in pairs.[1] Even if we choose not to discuss animals that aggregate for social or feeding purposes, we are left with a variety of clumped dispersions with no other basis for explanation than the avoidance of predation.

One of the most amazing manifestations of aggregation is the schooling of fishes, and biologists have for years explained schooling as an antipredator behavior. Schooling may have other roots, spawning or feeding in some cases, but clear and obvious antipredator benefits seem to be a necessary part of any explanation. Seghers (1974) compared schooling behaviors in stocks of the freshwater guppy, *Poecilia reticulata*, derived from different rivers in Trinidad and found that naive individuals from rivers containing many characid and cichlid predators exhibited well-developed schooling behaviors. Those stocks from rivers with little predation exhibited poorly-developed or no tendencies to school. Exposure to predators in the laboratory revealed greater vulnerability in the non-schooling stocks, although this result is difficult to attribute to schooling alone since guppies from heavily exploited populations exhibited a number of avoidance behaviors in addition to schooling.

One benefit that schooling does not provide is an effective group defense. The size difference between prey fish and their predators is usually so great that defense is impossible. Group wariness is enhanced certainly but is not as important in open water where prey cannot escape a determined pursuer (larger fish are faster than smaller fish). The primary advantage of schooling seems to be confusion of the predator when it attacks. The sheer volume of visual stimuli from a fleeing school is so great that it becomes nearly impossible to single out one prey individual for pursuit (Hobson, 1978). The consequence is a degree of protection for the individual that would not exist were it isolated.

The importance of predation to the social behavior of fish is clearly revealed by its comparison to an attractive alternate explanation of schooling – that it derives from hydrodynamic considerations. Weihs (1973, 1975) calculated that fish might profit energetically from several features of the movement of water within a school. A swimming fish leaves in its wake a series of spinning vortices (Fig. 9.1) which can either help or hinder followers, depending upon their position. A follower swims most easily by placing itself mid-way between the preceding two fish and at least five tail beats behind (the vortices do not stabilize immediately). Adjacent fish should be about 0.4 body lengths apart and beat their tails out of synchrony so that they can push off each other. The energetic savings from such behaviors, according to Weihs, are about five-fold – by no means a trivial economy. Obviously this explanation does not extend to benthic schools that move very little, but it might reasonably explain the phenomenon in pelagic species.

Some recent work by Partridge on school structure in saithe, herring and cod suggests that, for all its elegance, the hydrodynamic-efficiency hypothesis

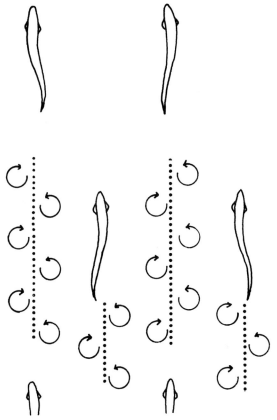

Figure 9.1   Fish placement and resultant hydrodynamics of a school swimming with minimum energy cost (after Partridge and Pitcher, 1979).

is probably wrong (Partridge and Pitcher, 1979; Partridge, 1980). The fish he observed do not swim sufficiently far behind those preceding them, nor are they centered. Adjacent fish are too far apart for optimal benefits and, even worse, are not usually on the same plane. Why do these species pursue what is an energetically suboptimal behavior? Partridge suggests that swimming on the same plane and very close to one's neighbors interferes with maneuverability and restricts the field of view, both of which are of paramount importance in the avoidance of predation. Here apparently is a circumstance where the energetic cost of antipredator behaviors can be calculated precisely.

When listing the possible benefits of aggregation, one cannot help but note that they fall into two broad categories. Some aspects of clumping benefit prey directly in their behavioral encounters with predators; others confer a more subtle advantage, usually involving some sort of interference with a predator's search.

Group defense provides the most obvious example of a direct benefit. If an ungulate such as a musk ox, an eland, or even a domestic cow can protect its sides and rump it is a formidable adversary; a group sufficiently large to allow the formation of a defensive circle permits a substantial reduction of vulnerability. Small birds which by themselves cannot deter a predator can mob it in groups and drive it away.[2]

Perhaps an even more important benefit of aggregation than group defense is an increase in wariness. Single individuals almost always succumb to surprise more easily than groups. A number of experimental and observational studies support this contention.[3] Aggregation can, in summary, both decrease vulnerability to attack and increase the time group members can devote to activities other than surveillance.

These are the obvious benefits of grouping. The more subtle idea that spatial aggregation might benefit prey indirectly through effects upon the predator's search became popular through papers by Rashevsky (1959) and Brock and Riffenburgh (1960). Their idea was quite simple: if a predator is searching an area containing ten fish, it is less likely to encounter anything if those ten are tightly clustered together than if they are spread apart. This line of reasoning follows the observation that each prey fish is surrounded by a circle of detection of radius equal to the perceptual distance of the predator with the deduction that if the circles overlap because the prey are clustered together, then the chance of a predator discovering anything as it moves through the area is reduced. Predators with no success in one locality may leave it entirely or may choose to switch to another species of prey. Such an advantage tends to be reduced if the predator, having discovered the grouped prey, gets to eat them all; so Brock and Riffenburgh attached to their searching model the constraint that, for various reasons, the predator may take only a fixed number from each school. The advantage to schooling accrues primarily to increases above this fixed number.

The suggestion that prey need have no group defense or group wariness to derive an advantage from schooling was appealing when first proposed but was somewhat at odds with the evidence. Ivlev (1961), working with fish, and Burnett (1958), using a parasitic wasp, had both observed greater success for a predator hunting sedentary and defenseless prey when those prey were clumped. Madden and Pimentel (1965) later observed the same effect with another parasitoid. The roots of this dilemma apparently lay in the theoretical assumption of random search employed by Brock and Riffenburgh. Unfortunately random searching has never been documented in any species. Ladybird beetles come as close to it as any type of predator, but even they accompany successful captures with a decrease in their speed and an increase in their rate of turning.[4] This seems to be a standard tactic of searching animals to keep them in the immediate vicinity of the last kill.[5] The ubiquity of this adaptation to clumped prey serves to make the Rashevsky–Brock and Riffenburgh theory suspect. As might be expected, subsequent theoretical work by Paloheimo (1971a,b) based upon more realistic models of searching has altered the original prediction. As far as the search is concerned, schooling of prey seems to benefit the predator's success rate rather than harm it. Limits on the number taken from each school must be imposed to counteract this advantage.

None of this theoretical work, not even Paloheimo's, mimics the searching behavior of predators sufficiently well to provide a reliable prediction of the effects of schooling. If a predator chooses to remain with a school for any length of time, then the necessary limit upon consumption becomes less and less of a constraint. On the basis of the experimental work I suggest that no indirect advantages of aggregation accrue to defenseless prey pursued by a more mobile predator.

Certain predator–prey systems do meet the assumptions of the Rashevsky–Brock and Riffenburgh model. Since that model founders on its failure to include a reasonable portrait of complex foraging behaviors, it might be expected to correspond more accurately to reality where such behaviors are relatively unimportant. The most common such circumstance occurs when the prey are much faster than the predator and the predator hunts from ambush. Ambush predators exhibit no complicated group-following behaviors; they simply attack an individual within range. Be the attack successful or not, other prey in the vicinity flee, breaking contact with the predator.

I have developed a model of this process which incorporates more reasonably than do earlier models limits to the rate of consumption of prey, and I offer a brief description of it here (Taylor, 1976, 1979). The model is a queuing process of the same general class presented in the last chapter but differing in several important details. Where the models in Chapter 8 required prey to be encountered at random, this one allows prey to aggregate into groups of size θ, such that with a mean density of H prey, groups of prey are encountered with frequency H/θ. An encounter alone is not enough to ensure a meal; prey must be successfully detected and attacked. The probability of detecting a group of θ prey, given its proximity, is $A(\theta)$. The probability of fruitfully attacking it is $B(\theta)$. The maximum mean rate of killing, therefore, is not H/θ but $A(\theta)B(\theta)(H/\theta)$. But the maximum rate is rarely realized; an attack on a detected group is forestalled many times because the predator is occupied with the consumption and digestion of a recently-captured victim, which activities it performs at mean rate μ. One can demonstrate that any advantage to the prey of aggregating when faced with such a predator declines with an increase in the ratio of the maximum killing rate to the rate of handling and digestion:

$$\rho = \frac{A(\theta)B(\theta)[H/\theta]}{\mu} \tag{9.1}$$

The benefits of group living, therefore, only exist in the circumstance when ρ

declines with increasing group size, or

$$\frac{d\rho}{d\theta} = \frac{H}{\mu\theta^2}\left[\theta\,\frac{d(AB)}{d\theta} - AB\right] < 0 \qquad (9.2)$$

The sign of the term in brackets determines the sign of the equation, so group benefits emerge when $d(AB)/d\theta < AB/\theta$, or

$$\frac{1}{A}\frac{dA}{d\theta} + \frac{1}{B}\frac{dB}{d\theta} < \frac{1}{\theta} \qquad (9.3)$$

This inequality can be made a bit easier to understand by use of the same graphical method Armstrong employed to investigate the functional response (Chapter 8). Figure 9.2 displays a hypothetical form of the function $A(\theta)$, the dependence of the probability of detection upon group size. At a given point on the abscissa $dA/d\theta = A/(\theta-a)$, where a is the θ-intercept of the tangent line.

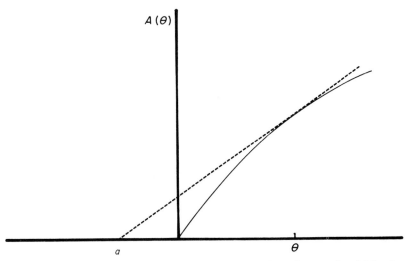

A (θ)

a

θ

Figure 9.2   The probability of detection, $A(\theta)$, as a function of group size, θ. The slope of the tangent line at θ is $A(\theta)/(\theta-a)$.

From this it follows that $(1/A)\,dA/d\theta = 1/(\theta-a)$. Following the same argument, the tangent to the function B will intersect the abscissa at point b. By substitution Inequality (9.3) becomes $1/(\theta-a)+1/(\theta-b)<1/\theta$, rearrangement of which yields the following (for $a<0$): if $b<0$, then prey benefit when $b<\theta^2/a$; if $b>0$, then prey benefit when $b>\theta^2/a$.

In this case, where A is an increasing but decelerating function of group size $(a<0)$; if prey are less vulnerable in groups $(b>0)$, then they always benefit from clumping. Even if prey are vulnerable to attack in larger groups $(b<0)$, then they can still benefit from clumping as long as group size is large enough to satisfy the relationship, $\theta^2/a > b$.

An interesting second case occurs when A is sigmoid $(0 < a < \theta)$, perhaps reflecting reliance upon a crypsis effective only with small groups. The conditions for benefit when prey are less vulnerable in groups $(b > \theta)$ now become quite stringent. Group size must fall within the very narrow range where $\theta < b < \theta^2/a$. On the other hand if prey are more vulnerable to attack in groups $(b < \theta)$, then, over the range of group sizes making a positive no benefits accrue from aggregating.[6]

To summarize, given ambush predators:

1. Where prey do not employ crypsis, they benefit from aggregation above a threshold group size regardless of the form of the function B.
2. Where prey use crypsis and B is an increasing function of group size, then aggregation is harmful when it contributes dramatically to the breakdown of crypsis.
3. Even when aggregation confers a decrease in vulnerability to attack, the clumping of cryptic prey must exceed a threshold to be of benefit, that threshold is the point at which the effects of loss of cryptic protection begin to stabilize.

One other circumstance can lead to indirect benefits from clumping, and this operates even when predators are more mobile than their prey. In Chapter 8 I suggested that the various sensory cues used to find prey should be ranked in terms of their value, value being the improvement in the rate of killing as a consequence of learning the proper use of a cue. Does clumping influence the probability that a valuable cue will be learned? Tinbergen *et al.* (1967) and Croze (1970), working with carrion crows, claimed that specific search images develop more quickly when prey are grouped close together. The idea is that learning occurs more readily if the stimulus-response trials follow in rapid sequence. But this assumes that the predator is near the prey in the first place. The ecologically more important question may not be how a carrion crow learns to find prey on a beach but why the crow is on the beach in the first place. If an environment is not homogeneous but comprises instead a variety of habitats, would it not be possible for prey to benefit from clumping if by so doing they interfered with the predator's learning of a cue to the habitat, a cue which could be more valuable than the cue to the location of the individual prey? Perhaps the increase in vulnerability resulting from the rapid learning of a search image by the predators within the prey habitat would be more than compensated for by the relatively few predators that learned to use that habitat. This may seem implausible, at first glance, but I performed a series of experiments with grasshopper mice which gave this result (Taylor, 1977). An arena was prepared which contained paralyzed mealworms (late-instar larvae of the beetle, *Tenebrio molitor*) hidden in sawdust in one of the two arrangements shown in Fig. 9.3. The top of the sawdust over the six black

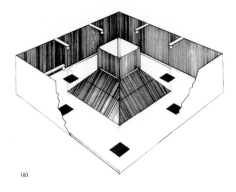

(a)

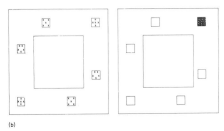

(b)

Figure 9.3 (a) Arena for observation of hunting grasshopper mice. The observer sat in the middle and watched the mice through slits in the blind. The inner wall of the arena was transparent acrylic plastic. (b) The two experimental arrangements of mealworms. The six squares were demarked by a light coating of lampblack on top of the sawdust. The small x's mark the locations of the individual prey (from Taylor, 1977).

squares was lightly dusted with lampblack to provide an initially neutral habitat cue. Mice were introduced into the arena and allowed to hunt for two hours. Those mice that were given dispersed prey to hunt fared substantially better than those hunting clumped mealworms. Mice hunting dispersed prey learned to use the habitat cue; those hunting clumped did not. This cue was sufficiently valuable that the rate of discovery in the dispersed arrangements was high. Had this been a natural environment, a female *Tenebrio* choosing an oviposition site might have profited by depositing her eggs in one spot rather than spreading them out.

The decision to aggregate or school with one's neighbors is an individual decision. To this point my arguments have been directed only to the level of the population. Williams (1966) and Hamilton (1971) maintained that prey might aggregate in the face of increased losses to predation for no other reason than individual selection for selfish behavior. The argument has been posed several ways but is similar and quite simple in each. Each prey animal seeks to minimize the area in which it is the nearest prey, reducing its area of vulnerability. It does this by moving next to another animal or, better yet, into the middle of a group of animals. Any subsequent confusion and vulnerability to attack are unimportant to an individual in the center of the group since predators tend to pick off stragglers and fringe individuals (e.g. Milinski, 1977a,b). This elegant argument was posed originally as a conflict between group and individual selection. The 'knocking off' of facile group-selection arguments was a popular and immensely satisfying intellectual sport of the late 1960s and early 1970s. The emphasis upon the benefit of the individual is productive, focusing attention as it does upon the internal structure and dynamics of the group, but such an approach does not require that prey suffer higher losses as groups. And in fact little evidence exists to suggest that they do.

What are the consequences of prey aggregation for population dynamics? This is an area of predation research that has not yet been adequately explored, but some tentative generalizations can be offered. If prey were sedentary and defenseless, normally one would expect predators to collect in areas where prey were most dense. The effect of this on the local level would clearly be stabilizing; the killing rate would be disproportionately high where prey were abundant and low where prey were sparse, generating a strongly sigmoid functional response (Royama, 1971; Hassell and May, 1974). Typically, when predators crowd into an area of high prey density they interfere with one another's hunting efforts. This would probably result in an additional degree of stability on the level of the metapopulation, although perhaps at the expense of a higher equilibrium density of prey. Beddington *et al.* (1978) claim that aggregative responses by predators and their interference with one another when aggregated are the two most important stabilizing influences in successful cases of the biological control of insect pests by parasitoids.[7] The extent to which their claims extend beyond the arthropod host–parasitoid world is not known, but if the logic they employ is correct we can expect stability if predators respond to aggregations and instability if they do not.

Even the supporters of this argument apply it only to the circumstance of actively-moving predators hunting sedentary, defenseless prey. If prey are more mobile than the predator and less vulnerable to attack in groups, then the dynamical influence of prey aggregation need no longer be stabilizing. A density increase resulting in larger groups of prey might actually produce a decline in a predator's rate of killing. What happens in this circumstance must depend heavily upon how patterns of aggregation vary with prey density. We know those patterns do change, sometimes dramatically (e.g. Rogers, 1977), and we can speculate on how those changes should depend upon density; beyond this, however, we cannot proceed. Either the empirical data do not exist, or else such data as do exist have not been analyzed with this purpose. In the case of ambush predators, a few tentative predictions of their population impact can be made by insertion of the ambush queuing model into the

Holling disc equation (Taylor, 1981)

$$f(H) = \frac{AB[H/\theta]}{1 + \dfrac{AB}{\mu}[H/\theta]} \tag{9.4}$$

The shape of this functional response reflects the dependence of detectability, $A(\theta)$, and attack success, $B(\theta)$, upon group size and, in turn, the dependence of group size upon density. If group size does not vary with density, then A, B and θ remain constant, and a typical type II functional response results. The complications arise if θ is a function of density. Consider two alternatives, where group size varies in proportion to density and where it is an accelerating function of density.

In the first case the density of groups of prey is fixed, such that an increase in numbers of prey results in larger groups, or $H = \beta\theta$, where β is the mean density of groups. Substitution into Equation (9.4) gives

$$f(H) = \frac{(AB)\beta}{1 + (AB)(\beta/\mu)} \tag{9.5}$$

The dependence of $f(H)$ upon density in this case mirrors the dependence of AB upon group size because

$$\frac{\mathrm{d}f(H)}{\mathrm{d}H} = \frac{\dfrac{\mathrm{d}(AB)}{\mathrm{d}\theta}}{[1 + (AB)(\beta/\mu)]^2} \tag{9.6}$$

If the product of A and B is a decreasing function of group size, then the slope of the functional response is negative. Only if the slope of the product is positive does the functional response display an increase with prey density. The biologically reasonable forms of $A(\theta)$ and $B(\theta)$ are those that result in prey in groups being more visible but less vulnerable to attack. The functional response curves that such forms of the two functions produce can be somewhat complicated. For the most part they have either zero or negative slope; the regions of positive slope are restricted.

The case for a stabilizing functional response deteriorates further when group size is a more sharply increasing function of density. Such might occur if prey rely upon crypsis when sparse, but that protection breaks down at higher densities. Prey might then change their social behavior and go to larger groups. If, for the sake of argument, $\theta = \alpha H^2$, then Equation (9.4) becomes

$$f(H) = \frac{\left(\dfrac{AB}{\alpha H}\right)}{1 + \left(\dfrac{AB}{\mu\alpha H}\right)} \tag{9.7}$$

The derivative of this function is positive only if

$$H \frac{d(AB)}{dH} - AB > 0 \tag{9.8}$$

Expressing this in terms of $d(AB)/d\theta$ and rearranging yields as the criterion for positive slope that $dAB/d\theta > AB/2\alpha H^2$, or substituting $\theta = \alpha H^2$

$$\frac{dAB}{d\theta} > \frac{AB}{2\theta} \tag{9.9}$$

This is a much more stringent condition than that emerging from Equation (9.6) and is likely to be met only if $B(\theta)$, the probability of successful attack, increases with group size in an accelerating fashion.[8] I must emphasize that the question to which these inequalities applies is whether the functional response curve has positive or negative slope, that $f(H)$ has a positive slope is not sufficient to guarantee that it is stabilizing.

Can one reasonably conclude that an increase in density can actually reduce the availability of prey? Several cases of a type IV or dome-shaped functional response have been described in the literature (Mori and Chant, 1966; Tostawaryk, 1972; Nelmes, 1974). Tostawaryk's work on the sawflies, *Neodiprion swainei* and *N. pratti banksianae*, and their pentatomid predator, *Podisus modestus*, is particularly interesting in this respect. At densities of 2 and 5 larvae per branch, the sawflies chose to disperse; at a density of 10 they formed a loose aggregate; and at densities of 20 and 40 they formed a compact cluster. Both sawfly species struggle vigorously when attacked and may exude a sticky resinous material from their mouthparts. This was rarely a serious deterrent to the predator's attacks upon solitary larvae, but attacks upon grouped larvae were another story. Some pentatomids came away from an assault upon a colony so covered with the sticky substance that they had difficulty moving, much less mounting further attacks. The result of this effective group defense was the dome-shaped functional response in Fig. 9.4.

Any phenomenon that causes a decline in the rate of predation per predator with increased prey density will clearly tend to destabilize a system. With ambush predators I suggest that stability can be enhanced by increased prey aggregation only if that clumping causes drastic increases in detectability and vulnerability to attack. Since mobile prey have sole control over their pattern of grouping, this condition is not likely to be met. In most circumstances, selection should operate against behaviors which increase vulnerability and, as a result, operate against stability.

I have argued as though the only aspect of the interaction to change with the density of the prey is the degree of prey aggregation. But Formanowicz (1982) has discovered that the dytiscid beetle, *Dytiscus verticalis*, alters its hunting behavior from active foraging when prey are sparse to ambush predation when prey are abundant. I suspect that many predators retain this kind of flexibility.

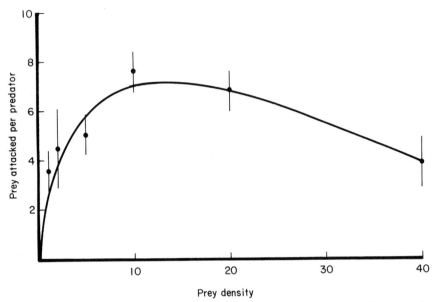

Figure 9.4   Functional response of second-instar *Podisus modestus* attacking second-instar *Neodiprion swainei*. Error bars are ± 1 SE (after Tostawaryk, 1972).

Akre and Johnson (1979) suggest such a shift as the only reasonable explanation for the switching they observed in naiads of the damselfly, *Anomalagria hastatum*. The dynamical consequences of this sort of density-dependent change in tactics could be interesting.

A certain amount of confusion is characteristic of any field in which research is still active. As a topic, the influence of the spatial structure of prey populations has more than its share perhaps. I suggest that this is, to a large extent, the consequence of an inadequate empirical foundation for theoretical speculation. Experimental work on this subject is difficult, but until more of it is done, most of the assertions in this chapter can only be considered tentative.

NOTES

1. See also Crook (1965), Horn (1968) and Thompson *et al.* (1974).
2. Crook (1964, 1965) and Horn (1968).
3. Crook (1964), Powell (1974), Page and Whitacre (1975), Siegfried and Underhill (1975) and Kenward (1978).
4. See, for example, Banks (1959), Bänsch (1966), Dixon (1959) and Fleschner (1950).
5. Curio (1976), Smith (1974, 1975) and Taylor (1977).

6. This point was incorrect in the original paper. I am grateful to Michael Cain for pointing out my error.

7. The extent to which one can hope to model an indefinite number of field systems with one set of two interacting equations ought to be considered when reading this and similar work.

8. The original paper posed this condition as $\mathrm{d}AB/\mathrm{d}\theta > AB/\theta$, which is incorrect. The precise criterion for an increasing $f(H)$ is $a + b < 2\theta - 1/2\theta$.

10 Predation and population cycles

Now the number of mice is largely dependent,
as everyone knows, on the number of cats.

Darwin (1859)

The first attempts at a mathematical theory of predation appeared in the 1920s, the well-known Lotka–Volterra equations. The major conclusion of this simple theory was that predator–prey systems tended to oscillate numerically with periods and amplitudes determined by their initial population densities. The coincidence of this theory with the first quantitative analyses of cyclic fluctuations of small mammal populations by Elton (1924) led to the durable hypothesis that such oscillations were manifestations of predation.

Although the Lotka–Volterra equations came under attack as early as the 1930s, they remained, in the minds of most ecologists, acceptable as capturing the essence of the predatory interaction. Subsequent empirical work on cyclic voles, lemmings and hares laid to rest meteorological explanations of this curious phenomenon, while leaving predation and other biological explanations unrejected (Elton and Nicholson, 1942). Eventually it was discovered that numbers of prey animals at peak densities vastly exceeded the potential regulatory capacity of local populations of predators, but this proved no barrier to the continued invocation of Lotka–Volterra cycles. Lack (1954) simply shifted the process one step down in the food chain, observing a predator–prey cycle between herbivores and their food plants.

The advent of modern predator–prey theory in the 1960s raised a sobering issue. According to this theory (summarized in Chapter 2) the Lotka–Volterra equations (generating as they do a neutral equilibrium, neither stable nor unstable) are a special case, infinitely unlikely to occur. The neutral oscillations which they predict disappear with the first attempt to make the model more realistic, a phenomenon mathematicians call structural instability. This posed a logical problem of the first order, but before it had a chance to penetrate very far into the ecological mind, limit cycles made their appearance in the literature. Lecture notes were quickly changed from neutral

Lotka–Volterra cycles to stable limit cycles, and predation continued to be invoked as a cause of population cycles with scarcely a pause.

The question motivating this chapter is a simple one: is it true that natural oscillations of populations are driven by predation? This can certainly be true in theory; moreover, the concept of a limit cycle is appealing on esthetic grounds. Because limit cycles have their periods and amplitudes determined intrinsically by the rates of the various processes that generate them, not extrinsically by the initial starting conditions, they are stable to perturbation. This is a necessary feature for persistence of an oscillation in a stochastic environment. Of course the general theory need not be precisely descriptive of the field situation, so we should not expect neat deterministic limit cycles in nature. Perhaps the question should be rephrased to ask if the theory is sufficiently robust in its qualitative predictions of stable cycles to serve as a general guide to naturally cyclic populations.

The first task in examining the possible role of predation in animal population cycles is to define more precisely what is meant by the word *cycle*. Trajectories of populations in the field are not cyclic in the strict sense of fitting an oscillatory mathematical function (Garsd and Howard, 1981). They are certainly characterized by alternate periods of increase followed by periods of decline, but the sizes of the peak populations and, to a lesser extent, the intervals between peaks are always variable. To the layman, however, periodic outbreaks of insects or plagues of voles are cycles, and so cycles we call them. However the fluctuations are described, they are dramatic, difficult to explain and, therefore, deserving of attention.

Before looking at case histories of natural populations it might be reasonable to ask whether predation-induced limit cycles have been found in laboratory systems. If one cannot produce them in carefully controlled conditions in the laboratory then one probably need not agonize over their existence in the field, where simple models are much less likely to apply. The best laboratory evidence comes from continuous cultures of microorganisms in chemostats. Jost *et al.* (1973a) presented the results of a series of chemostat experiments using as prey the bacterium *Azotobacter vinelandii* grown in a medium featuring glucose as the energy source. The predator upon the bacterium was the ciliate protozoan *Tetrahymena pyriformis*. As I mentioned in Chapter 5, the operator of such a chemostat has control over two variables, the concentration of glucose in the inflow and the rate at which the medium is replenished. By suitable manipulation of these two variables, the authors were able to discover continued stable oscillations in the predator and prey (Fig. 10.1); moreover, a mathematical model of this system predicts limit cycles at precisely those sets of the control variables that produced cycles in the experiments (Fig. 10.2).[1]

One can only conclude that limit cycles are indeed possible in predator–prey systems and that a look at natural systems appears to be justified. I shall discuss three kinds of oscillations in which predation may be important,

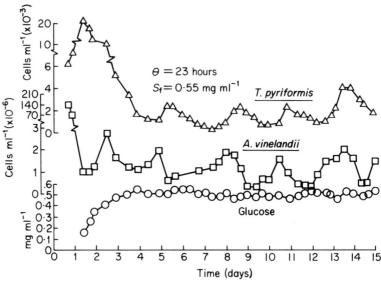

Figure 10.1   Stable cycles in a chemostat culture of *Tetrahymena pyriformis* feeding upon *Azotobacter vinelandii* (from Jost *et al.*, 1973a).

outbreak patterns in two insects, the 3–5 year cycle of microtine rodents, and the ten-year cycle of snowshoe hares.

The spruce budworm is a serious forest pest in Canada. Typically an outbreak species, it punctuates long periods of relative scarcity with massive eruptions of larvae. The result is the defoliation, and in many cases the death, of host trees. Morris *et al.* (1958) reported on a ten-year study of the numerical responses of avian and mammalian predators to changes in budworm density during outbreaks. Several species of insectivorous birds responded to outbreaks with striking increases in the densities of nesting pairs; the bay-breasted warbler, in particular, increased to 12 times its previous density. The budworm densities, however, increased by a factor of 8000. The estimated total predation by birds, less than one percent, was clearly insufficient to forestall further increase. The limit on budworm numbers was set by depletion of the foliage upon which they fed, not by predation. Morris and his coworkers concluded with, to my knowledge, the first statement of a hypothesis which has subsequently become popular. Avian predation, they claimed, is insufficient to stop an outbreak once it starts, but it may prove quite adequate to maintain the budworm at low densities between outbreaks. In other words, predation may serve the functions of hastening declines and lengthening the periods between outbreaks.

Dixon (1971) and Wratten (1973) provide another example of an insect oscillation with their research on the lime aphid and its predators. Lime aphids show periods of rapid population growth, which cease not because of

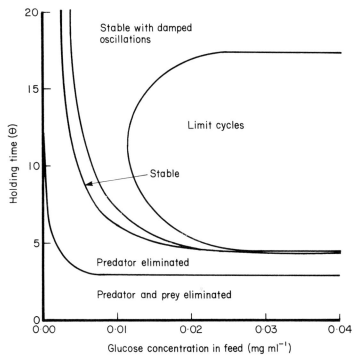

Figure 10.2   Influence of holding time and glucose concentration in the feed upon the dynamics of a model of the *T. pyriformis–A. vinelandii* interaction (after Jost *et al.*, 1973b).

increasing mortality but because of intraspecific processes that result in stunted females of low fecundity. Their primary enemy, the coccinellid beetle *Adalia bipunctata*, chooses to oviposit on trees which already have high densities of aphids. Beetle larvae, as a consequence, begin to influence aphid numbers far too late to forestall the increase phase. The voracious later instars of the coccinellid coincide with declining numbers of lime aphids and drive prey numbers quite low, lower certainly than would be expected on the basis of the crowding effects that the aphids produce themselves. As a result, lime aphids in the presence of *Adalia* fluctuate rather more dramatically than one would expect in their absence. As with the spruce budworm, the dynamics of this insect may reflect predation, but predation is clearly not the whole story. Neither system suggests a simple two-species limit cycle.

Shifting perspective a bit, I examine next a classic problem in population ecology, the cycling of voles and lemmings. The most facile explanation of microtine cycling is that it is the result of an indissoluble complex of factors, both environmental and internal to a population, which are likely to vary in importance according to local conditions. As Chitty and Krebs have pointed out, such an explanation is safe to the point of tedium, in that its rejection is virtually impossible. They suggested that the preparation of hypotheses which are scientific, rather than metaphysical, requires that one first step out onto a limb and posit one underlying explanation of cycles in all populations of the approximately 100 microtine species.[2] This is a breathtaking assertion, clearly wrong at some level of quantitative prediction, but refreshing nonetheless. The continuing vigor of research in this field can be traced directly to the willingness of its pioneers to sustain criticism. Chitty wrote in 1967 that 'perhaps ecologists of the next generation will be more successful. . . . They might start by doubting the truth of everything that has so far been written on the subject, including the ideas of the present reviewer.'[3]

A number of studies have followed predation through portions of a cycle. Of these the most well-known is Pearson's work on *Microtus californicus* (Pearson, 1966, 1971), in which he correlated heavy mammalian predation with decreases in the density of a vole population. Pearson suggested that while predation cannot account for the cessation of population growth it may drive a static population down and hold it at low densities, setting thereby the period of the cycle. This is a mechanism similar to that proposed by Morris for insect populations. Pearson concentrated upon mammalian predators, but one might expect avian predators to exert a similar influence.

This idea has not been well received by the advocates of the dominant theory of microtine cycling, the behavioral-genetic mechanism. This hypothesis accounts for oscillations in vole numbers by qualitative changes in the structure of a population. A polymorphism in genetically-determined behavioral types is supposed to drive the cycle by alternation between greater fitness at high densities of a behaviorally aggressive form which breeds poorly, and at low densities of a fast breeding and more socially tolerant animal. Such a mechanism can account for much of the current data on microtine dynamics and has the virtue that it is yet to be definitively disproved. The most succinct recent statement of this hypothesis is provided by Krebs (1978b).

Its advocates dislike Pearson's, or any other, predation hypothesis for two reasons. The first is that heavy predation pressure is not always observed to be coincident with vole declines. This is a difficult criticism to take seriously. The intensity of predation, especially nocturnal predation, is remarkably difficult to estimate by casual observation. Even as inadequate and indirect as scat and pellet analyses are, they require a great deal of work and a sophisticated knowledge of the behaviors of local predators. The sorts of methods used by Schnell (1968) to account for predatory losses from a fenced population of cotton rats are usually necessary. To find a significant percentage of the individuals killed by birds of prey, he inserted radioactive cobalt or zinc pins under the skins of all cotton rats in the fenced population and then searched with a Geiger counter beneath roosting sites of local hawks and owls. But even if one accepts casual natural history observations as adequate descriptors of the intensity of predation, the fact remains that microtine cycles are not well described. Enormous variations occur from place to place, or even in the same location at different times, in the sizes of the peaks and the intervals between them.[4] Until the causes of such variations in the form of the cycle are better understood, it is premature to assert that the apparent lack of predators has no effect.

The second objection is that 'the role of predation in microtine cycles is limited to the mortality component of the demographic machinery, and consequently other factors must be invoked to explain reproductive or growth changes' (Krebs and Myers, 1974). On this evidence alone microtine dynamics cannot be explained as a simple predator–prey limit cycle. Since a wealth of evidence exists for reproductive, physiological and other changes in the qualities of the animals in a population over the course of a cycle, we are required, if we admit to any role for predation, to allow for the action of two separate processes, one intrinsic and one extrinsic. The advocates of the behavioral–genetic hypothesis claim to need only one mechanism, that being intrinsic. It is not clear that this is true.

The current form of their argument relies a great deal upon the process of dispersal of voles away from expanding and dense populations. If these emigrants are genetically atypical of the source population, that population will change in character as a consequence of the exodus. For this argument to work, the dispersers must disappear into a 'sink' in the surrounding habitat, otherwise one population's loss must necessarily be an adjacent one's gain (Tamarin, 1977, 1978). The only reasonable source of such a sink is a community of predators waiting in the wings. Unless another mechanism can be found for the summary destruction of dispersing voles, I find it unclear that the behavioral–genetic hypothesis can stand alone as a sufficient explanation of microtine dynamics.

But let us assume for the sake of argument that it can. The criticism its advocates then level at predation and other extrinsic processes is that they are less parsimonious explanations of microtine fluctuations. If so, then the Popperian approach would have us put them on a back burner until the more simple, intrinsic theories had been adequately researched and rejected (Popper, 1959).[5] This response, I suggest, may be a misreading of Occam's Razor. The criterion of simplicity (which by Popper's definition equates to ease of falsification) is practical, not abstract. I would be hard pressed to defend the assertion that every hypothesis invoking an intrinsic mechanism is cheaper, faster, and easier to falsify than those which employ extrinsic mechanisms. If, for example, the genetic or demographic structure of natural microtine populations proved to be more difficult to alter than levels of predation, then theories whose tests required genetic or demographic manipulation would, by definition, be less simple.

All of this is not to claim that predation is critical to microtine dynamics, although it has not yet been shown to be unimportant. I claim only that the question of the extent to which microtine cycles would be altered by removal or addition of predators remains an important one and hope that the short-term exclusion studies reported by Taitt and Krebs (1983) on *Microtus townsendii* will be the first of many to examine this subject experimentally.

The lynx–snowshoe hare cycle in North America comprises the third of the case histories of oscillatory populations to be examined here. Elton and Nicholson (1942) summarized an enormous volume of data from trapping records which implied a ten-year cycle in lynx (Fig. 10.3). The reality of this

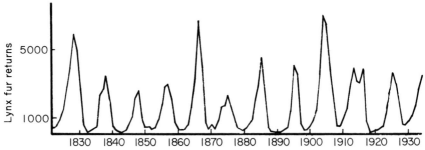

Figure 10.3 Lynx fur returns from the MacKenzie River region of Canada (from Elton and Nicholson, 1942).

cycle has been examined on various occasions subsequently and adequately defended (Keith, 1963; Finerty, 1980).[6] As popular as this system is, remarkably little field research has been done on it. Most of it has come from Keith and his students, and I shall report their version of the cycle (Keith, 1974; Keith and Windberg, 1978). Figure 10.4 shows the change in numbers of hares on four study areas in central Alberta from 1966 to 1975. The highest densities (April 1971) were about 29-times greater than the low in the spring of 1966. When available, the snowshoe hare provides the bulk of the diet of lynx, coyote, red fox, fisher, and two important avian predators, the great horned owl and the goshawk. The lynx seems to be the most specialized upon hares, although the others undoubtedly comprise an important source of mortality. During the upswing of the cycle, hares increase at a rate which exceeds the predatory capacity of their enemies. Cessation of population growth, as a consequence, comes not from predation but from the depletion of the shrubby vegetation upon which hares rely for winter browse. Hares have difficulty maintaining weight over the winter, but the average weight loss in the winter immediately preceding the population peak was substantially higher than normal and continued high for several years. Correlated with this, the productivity of each female hare began to drop in the summer of the peak and continued to decline for 2 or 3 years thereafter. Juveniles grew more slowly during the decline. At about this point in the cycle, predators, which had been reproducing well for a couple of years, began to exert an influence. The winter survival of adult and juvenile hares declined, particularly that of the juveniles.

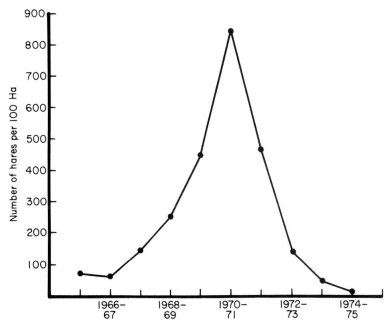

Figure 10.4   One cycle in showshoe hare abundance in the Rochester district of Alberta (after Keith and Windberg, 1978).

By the mid-1970s the predators had driven hares to quite low densities. During this trough in the population cycle, winter weight loss declined, hares reproduced better, and juvenile growth and survival improved. But numbers did not increase because of continued heavy predation. Eventually the raptorial birds decreased in abundance; they either failed to breed or left the areas where hares were sparse.[7] Lynx experienced nearly complete reproductive failure at this point; litter sizes were the same, but most kittens died before reaching maturity. Mortality of adult lynx remained low, interestingly enough; apparently adults are efficient enough hunters to survive even when food is scarce. But an eventual decline in predators is inevitable and sets the stage for the next period of explosive increase in the hare population.

An interesting phenomenon accompanying the lynx–hare cycle is a synchronous cycle in some upland game birds. The ruffed grouse increases during peak hare years and declines when hares are sparse. Evidence suggests that this results from the transfer of attention of raptorial birds to hares when they are available and to grouse when they are not. Rusch *et al.* (1972) document such a change in the diets of great horned owls during a five-fold increase in hare densities.

The three-level hypothesis is clearly becoming an attractive explanation to a number of wildlife ecologists. The idea that predators do not stop population increase in their prey but serve to drive declining populations to low densities seems to fit the evidence for a number of field systems, in addition to that of the hare and its predators. It correlates quite nicely with the dynamics exhibited by wolves and moose on Isle Royale, discussed in Chapter 4. It seems to fit what is known of coyote–jackrabbit interactions in the American West (Wagner and Stoddart, 1972).

I offer a caution, however. As has happened with any number of other ecological hypotheses, this idea is being elevated to the status of fact without adequate analysis. The logic itself is still in the process of development, and a great deal of uncertainty exists over the source of the oscillation. Several possibilities offer themselves for the dynamics of a plant–herbivore–predator system. The first is that the plant–herbivore interaction is, by its own properties, oscillatory with a period and amplitude of the sort observed in the field, and predator populations are simply entrained into this cycle by virtue of their dependence upon the herbivore for food. This is similar to Lack's hypothesis for population cycles (although he postulated Lotka–Volterra cycles instead of limit cycles as the driving force) and, with Lack, requires the acceptance of predation as a nearly insignificant force in the dynamics of oscillatory herbivores.

The second class of dynamic explanations is that plants and their herbivores will oscillate alone. Predators neither create nor destroy these cycles, but they do fundamentally alter their nature. The alteration typically suggested is a lengthening of the period of a cycle.

The third possible explanation is that cycles are a property only of the three-species interaction. Neither the herbivore–plant nor the predator–herbivore subsystem need oscillate when isolated from the third species, but all three together generate a cycle with the appropriate properties.

How are we to distinguish among the three? Keith advocates the second explanation. Finerty (1980), whose well-written and entertaining evaluation of population cycles is probably the most complete and statistically respectable analysis of the data extant, agrees with him:

> One can expect that many of the causal factors producing cycles in prey will also affect predators, but it seems likely that the cycle is not being generated by interactions between vertebrate predator and prey. (P. 102.)

Both are convinced that predation shapes the cycle but does not originate it by the fact that hares at their peak exhibit clear evidence of limitation by food. Although motivated by a different theory, Krebs and Myers (1974) apply the same sort of logic to voles when they assert that predation could not possibly account for the qualitative changes observed in peak and declining populations.

I suggest that this is short-sighted reasoning. Documentation of a physiological or reproductive change in a herbivore indicates quite clearly that an important alteration in the environment has occurred, be it a change in food supply or quality, or a change in the genetic–behavioral properties of the herbivore population. But it tells little about the source of that environmental change. Is it not equally plausible that hares increase to such high densities because predation has suppressed them for so long that their food supply has grown to enormous levels? Were predators not present, hares might prevent winter browse from attaining levels of abundance that would support such explosive growth in their numbers. The indirect influence of predators upon the food supply of the prey must be acknowledged if one invokes a three-level explanation of cycling.

Finerty offered a related line of reasoning in support of the first and second hypotheses and against the third. He argued first, and quite correctly, that a time-lag in a variety of simple models of single-species growth can produce limit cycles (May, 1976). And second, such a time lag could result from a delay in the recovery of vegetation from browsing. Then he concluded, again accurately, that predator populations could be forced into an oscillation by an independent cycle of the herbivore and its plants. Hypotheses 1 and 2, in other words, find support in mathematical theory.

What Finerty failed to mention is that the third hypothesis is also theoretically sound. A small body of mathematical literature exists on three-level food chains,[8] and it has produced a number of predictions. Two are of interest here. If a herbivore and its food supply exhibit a limit cycle, then the entrance of a carnivore into the system can result in its destabilization, in the continuation of cycling, or in the production of a stable equilibrium. Likewise, if the herbivore population is in equilibrium with its food supply, the introduction of a predator may destroy that stability, enhance it, or produce a limit cycle. Which of these possibilities occurs depends upon various features of the biology of the species in the food chain, features which for the most part are not well understood in the field studies under examination. The conclusion is inescapable. Prediction of the dynamics of a three-level food chain cannot be made from an understanding of either of the two subordinate trophic interactions.

Discrimination among these three hypotheses requires experimental removal of predators and documentation of the resultant dynamics of the herbivores. Such experiments may be difficult, financially impractical, or impossible in some circumstances. But without them these and other alternative hypotheses must remain tentative and untested.

NOTES

1. The model is of much the same form as Equations (5.1)–(5.3).
2. Chitty (1960, 1976a), Krebs and Myers (1974) and Krebs (1978b).
3. Chitty (1967b) quoted in Krebs (1978).
4. The time between population peaks averages 3 to 4 years. Longer intervals are not uncommon, and shorter 'cycles' of 2 or even one year have been

described (Lidicker, 1973; Krebs and Myers, 1974; Birney *et al.*, 1976; Abramsky and Tracy, 1979).

5. A more digestible summary of his argument may be found in Hempel (1966).

6. That the snowshoe hare cycle involves predators is well established. A couple of decades ago, Keith (1963) cast doubt on this hypothesis by reporting that hares oscillate on Anticosti Island in the absence of lynx. This was impressive evidence to some of the irrelevance of predation, e.g.

> ... at a stroke, Keith (1963) was able to refute the hypothesis that snowshoe hare cycles are part of an intrinsic predator–prey oscillation when he found that ... (Simberloff, 1983)

Finerty (1980) demolished this counterexample by pointing out that Anticosti Island was not predator-free; foxes were so abundant that trappers harvested thousands of pelts from the Island.

7. This mass exodus occurs during the winter and apparently accounts for the periodic influxes of Canadian hawks and owls into the northern portions of the United States.

8. Rosenzweig (1973a), Wollkind (1976) and Lin and Kahn (1977).

11 The evolution of predator–prey systems

> Of a truth Divine Providence does appear to
> be, as indeed one might expect beforehand, a
> wise contriver. For timid animals which are a
> prey to others are all made to produce young
> abundantly, that so the species may not be
> entirely eaten up and lost; while savage and
> noxious creatures are made very unfruitful.
>
> *Herodotus*[1]

Ecological systems *per se* do not evolve; their constituent species' populations evolve. The effects of evolution upon the dynamics of predatory systems must, therefore, be sought in the selective forces operating separately upon the predator and prey. Books have been written about little else but evolutionary adaptations for and against predation. Protective morphology, coloration and behavior provide a substantial fraction of the stories with which ecologists regularly amaze one another. Much of the systematic work on predators employs teeth, jaws, raptorial appendages, and other adaptations for pursuit and capture.

Here I shall take these things as relatively fixed, albeit important, attributes of the organisms in a particular system and concentrate instead upon more genetically plastic adaptations. The goal in this chapter is not to come up with grand pronouncements about why predator–prey systems are as they are, but rather to predict how they might change over a short time scale. I shall attempt, in pursuit of this goal, to wend my way carefully between the Scylla of assuming genetic fixity in ecological time and the Charybdis of assuming ecological stasis in evolutionary time.[2]

Of the various ideas about predation that ecologists have taken to heart over the years, one of the most durable is that predators are necessary for the genetic health of their prey populations. That 'genetic health' is a difficult term to define bothers remarkably few in the face of substantial evidence that predators direct their efforts toward the young, elderly, ill and injured among their prey.[3] If, in fact, these subsets of the prey population differ genetically from the whole, then it has not, to my knowledge, ever been demonstrated. Age and environmental quality during development probably account for the vast majority of these relatively more vulnerable prey.

Of course, predation does not fall randomly upon a prey population, and as a potential selective force its effects ought to be examined. As a point of departure it should be made clear that that which is selected for by predators is simply resistance to predation. The coincidence of this resistance with our conception of health or vigor is secondary, and in some cases the coincidence fails. Bergerud's study of lynx predation on caribou (Chapter 4) revealed that selection operates against aggressiveness, movement and curiosity in calves – qualities not usually taken to mean ill health or lack of vigor, but which in this circumstance take a calf away from its mother and bring it near ambush sites.

Prey and predator are involved in an evolutionary race, the prey to become more difficult to catch and eat, the predator to perfect its powers of search, pursuit, capture and killing. The task of the evolutionary biologist is to discover whether this game results in a coevolutionary stalemate or whether one side or the other eventually wins. The extinction of the loser (or, if the predator wins, both the winner and the loser), would imply that natural selection destabilizes predator–prey systems, a painful result to some, but one we must accept as possible. If the group selection debates of the late 1960s and early 1970s have taught us one thing, it is first to demand workable evolutionary mechanisms and then to let those mechanisms lead us to conclusions about what are or are not the results of natural selection. Bald assertions about those results, with no mechanism, are meaningless teleology.

Three distinct mechanisms have been proposed. The first, and most startling, is an explicit group selection model. The second is the 'prudent predator' concept. The third is the coevolutionary steady state model. The latter two have generated a great deal of confusion among ecologists.

Gilpin (1975) developed a portrait of a community of nearly-isolated predator–prey subsystems, in each of which the predator wins the evolutionary race by individual selection and drives the prey to extinction. The interesting aspect of his theory is that, on a broad scale, group selection prevails over individual selection; the more voracious phenotype dominates locally but loses to the less voracious morph over the whole complex of subsystems.

The traditional difficulty with group selection explanations is that their advocates have never been successful in coming up with plausible reasons for the high rates of extinction needed to make such mechanisms dominate individual selection. Gilpin produced a reasonable argument for extinctions by taking advantage of a feature of most simple predator–prey models, such as those in Chapter 2; they can become instantly unstable if the predator's voracity is increased.

The theoretical constraints on isolation of the local subsystems remains fairly strict, so Gilpin's theory may not be an appropriate paradigm for more than a small fraction of real-world systems. But the idea is intriguing nonetheless and deserves an experimental test.

The prudent predator concept was originated by Slobodkin and firmly woven into the intellectual fabric of ecology in his book published in 1961. Its premises are simple. Prey individuals can be harvested from a population at any rate, but only the subset of harvest rates that allows the population to maintain itself comprises that population's sustainable yields. By proper choice of the number and kind of individuals removed, the population yield can be maximized. The idea of a maximum sustainable yield grew out of the logistic equation and was popular in the 1960s as a paradigm for optimal human harvest of fish and game populations. Slobodkin thought it reasonable that natural selection should also favor a selectivity and intensity of predation that would maximize the yield of prey. Such predators would be acting prudently.

As originally posed, the argument was blatantly group selectionist, with only a hint of a mechanism.[4] In Slobodkin's defense, the concept was developed before Wynne-Edwards (1962) forced the issue of group selection, the backlash to which exposed such faulty logic in nearly everyone's thinking. When this aspect of his theory was pointed out (Maynard Smith and Slatkin, 1973; Ricklefs, 1973), Slobodkin chose to be combative. The resulting brouhaha raised some valuable issues and is worth recapitulating.

The unique features of Slobodkin's response in 1974 derived from arguments in the field of life-history evolution. He claimed that the disproportionate representation of young and old prey in a predator's diet means that the predator is harvesting optimally. To follow this argument, two concepts need to be mentioned. The first is that evolution must necessarily result in variation in age-specific vulnerability. The reason is that vulnerability arising from age-specific deleterious genes is most likely to be modified or eliminated in young animals, less quickly eliminated in older animals, and not eliminated at all in post-reproductives. The expression of genetic faults, in other words, will be directly correlated with age (Hamilton, 1966); therefore, we should expect elderly animals to comprise a proportionally greater fraction of a predator's diet. Not all vulnerability comes from bad genes. That of young animals probably arises from constraints upon development rather than from genetics. No matter how heavily parents invest in offspring, quite clear limits exist to the precocity of young animals. A certain amount of post-natal or post-hatching development must occur in any species before its young stand a chance against predators.

The second concept necessary to Slobodkin's argument is reproductive value, an idea originating with Fisher (1930). It is a measure of the ultimate worth of an individual of a given age to future reproduction in an exponentially growing population. Typically reproductive value is highest at the onset of sexual maturity, relatively lower at birth, and zero when an animal is post-reproductive. From the definition of the term alone, one can infer that the productivity of a population is least harmed by removal of those individuals with lowest reproductive value.

Slobodkin used these two concepts to deduce that selection will operate to shift reproductive value from those age classes of the prey most vulnerable to predation to the less vulnerable ages. Predators will simply choose the easiest animals to kill, and natural selection will operate so as to make these the prey with lowest reproductive value, those least important to the future growth of their population. The choice of victims, in other words, is not the result of predators consciously manipulating the productivity of their prey, as prudence would imply, but of natural selection operating upon the prey's life history.[5]

This argument has a fundamental flaw. If predators are restricted to killing prey animals of low reproductive value, then they lose control over the most critical feature of an optimal harvest policy, the reduction of the prey population below densities at which intraspecific competition interferes with productivity (Maiorana, 1976). At maximum densities a prey population has essentially no yield.

For the purposes of this book, the critical feature of the prudent-predator concept is that it does not produce population regulation. If predators take only non-reproductive individuals likely to die soon of other causes, they cannot prevent population increase (Kruuk, 1970). A fundamental problem with application of concepts from linear theory, such as reproductive value, is that they cannot be expected to apply when survivorship and fecundity both depend upon density and time (Mertz and Wade, 1976). For example, Mech's original data on the wolves and moose of Isle Royale show classic selection of old and infirm prey when wolves were few and moose relatively abundant. Subsequently, with higher wolf to moose ratios and a series of bad winters, this pattern changed. Before 1970 only 13% of moose killed were of 1–6 years in age; from 1970 to 1974 that proportion rose to 53% (Peterson, 1976).

There is an object lesson here. Slobodkin had a valuable point to make about harvesting optimally; differences in age and vigor among animals must be considered when deciding how best to manage a resource population for maximum yield. Unfortunately, in the making of it, he succumbed to the temptation to pander to the well-known penchant of ecologists for two-word theories. As a consequence, we must read the etymologically embarrassing claim of a well-respected biologist that although some species are born prudent and others achieve prudence, predators have prudence thrust upon them. Those of us in the next generation of ecologists had best think twice about relying upon catchy titles, if we are not also to be hoist with our own alliterative petards.

The third conceptual portrait of evolution in predator–prey systems, the coevolutionary steady-state, arose in the 1960s and early 1970s. The logical basis of this argument is the observation that rates of coevolution in predators and their prey do not remain constant but vary with the state of the system.

When prey are at the limits of their food supply and compete intensively with their own kind, predation will comprise a trivial portion of their selective environment. By contrast, sparse prey populations facing heavy mortality from predators will be strongly selected for mechanisms of defense. The consequences of natural selection operating in this fashion can be simultaneous coevolutionary and population steady-states.

This concept developed in a curious fashion. The earliest work on the subject was conducted by Pimentel and his students and was strictly empirical.[6] I review a portion of this work later in this chapter; suffice it to say here that he and his students discovered genetic changes in laboratory host–parasitoid systems that altered the stability of the interaction.

Apparently, Pimentel could not account for his interesting experimental results with any sort of coherent explanation, because he immediately began to generate a verbal haze that served to confuse ecologists for at least a decade. At one point he would assert that something called 'genetic feedback' regulated predator and prey densities. By the traditional use of feedback, this would imply that a population disturbed from equilibrium returns because it quickly changes in genetic composition. Such an argument poses an unrealistic time frame for genetic change. At another point he would invoke evolution as forcing predators to take only the 'interest' from a prey population and not its 'capital', an argument redolent of group selection. In a passage reminiscent of Slobodkin's prudent-predation arguments, Chabora and Pimentel (1970) asserted that:

> selection would generally eliminate populations which tend to fluctuate from an extremely high level to a low of a few individuals. Hence, the best strategy for parasite–host populations is integration and stability. Ideally, the highest efficiency of parasitism in terms of the maximum energy transferred from the host population would be at that point where sufficient hosts survive to successfully breed and replace the exploited individuals. Maximum host output for the parasite would be at the point where maximum harvest is possible with the resources available to the host population.

In Pimentel's defense, if one will publish empirical results, one must make at least a stab at explaining them, and Pimentel and his students have produced a wealth of data worthy of publication. But for an explanation of the coevolutionary steady state, we must turn to others.

Excavation of the logical roots of this proposed steady state have been approached in two different ways, one emphasizing genetic change in the interacting populations, the other phenotypic change. The approach specifying genetic changes will be discussed first, although the phenotypic approach has probably taken us farther. Explicit genetic models can allow for greater generality in the nature of the interaction between predator and prey, but they usually require simplistic assumptions about the mechanism of genetic change (one locus, two alleles). Models of phenotypic change make no assumptions about the nature of genetic change (which has bad as well as good aspects) but must in many cases be limited to a particular, specific form of the interaction Equations (2.7) and (2.8).

Levin and Udovic (1977) focus the difficulty of explicitly including genetic considerations in predator–prey models by pointing to six processes that must be considered in a genotypic model (Fig. 11.1). Loops 1 and 2 represent the only features of the predatory interaction discussed in the book until this

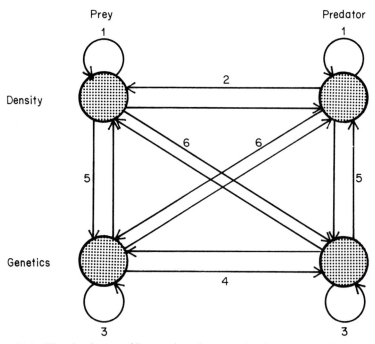

Figure 11.1   The six classes of interactions between density and genetics that must influence the evolution of predator–prey interactions (see text) (after Levin and Udovic, 1977).

chapter, namely the dependence of a population's growth upon its own density (loop 1) and upon the density of its predator or prey (loop 2). Recognition of the genetic structure of both populations adds four new classes of interactions: intraspecific density independent selection (loop 3), density independent interactions between genetic variables (loop 4), interactions between changes in density and genetic structure within each population (loop 5) and between each population (loop 6). Not surprisingly, models incorporating all six kinds of process are few in number and difficult to work with.

I shall not describe any such models in detail, other than to note a few general results from Levin and Udovic. They claim that in the circumstance where selection is frequency independent, that is where selection coefficients in a model do not depend upon the genetic structure of the populations, then a coevolutionary steady state can arise, for which there are three conditions: (1) the forms of the equations describing dynamics in numbers alone (e.g. Equations (2.7) and (2.8)) must ensure a stable equilibrium; (2) the prey population must maintain a balanced polymorphism by means of superior fitness of the heterozygote; (3) and so must the predator. These three conditions are independent, which means that genetic stability requires population stability; but other than through its influence upon parameter values, the genetic structure of the system does not alter population stability.

Note that this is a rather weak statement by comparison to Pimentel's assertions. He would have us believe that unstable systems can be stabilized by evolution. Levin and Udovic claim only that a stable system might not be destabilized by selection.

The assumption of frequency-independent selection is quite restrictive, requiring acceptance of selection coefficients in one population depending neither upon that population's own genetic structure nor upon the genetic structure of the population with which it interacts. Levin and Udovic note that frequency-dependent selection greatly increases the number of possible behaviors of a model. For example, a stable population model may be rendered unstable by consideration of genetic structure, or an unstable population model may be stabilized. And heterozygote superiority is no longer required for a balanced polymorphism. Perhaps this expansion of possibilities should be expected – models with four variables are bound to be more complicated than those with two. One can reasonably infer from this work and that of Levin (1972) that a coevolutionary steady state is reasonable, but it is not yet clear what kinds of things influence it.

Rosenzweig (1973b) and Schaffer and Rosenzweig (1978) are primarily responsible for the phenotypic approach. The idea is to try to isolate those contributions to the shapes and positions of the isoclines in the phase-plane portrait which are due only to coevolution in the predator and prey. The first step in doing this is to insert specific forms for the functions g, f and k into Equations (2.7) and (2.8). The forms chosen are the logistic equation, the disc equation, and a linear food-dependent growth equation,

$$g(H) = r(1 - H/K) \tag{11.1}$$

$$f(H, P) = \frac{aHP}{1 + ahH} \tag{11.2}$$

$$K[P, f(H, P)] = P\left[\frac{\beta aH}{1 + ahH} - d\right] \tag{11.3}$$

Of the two unfamiliar parameters, d is the death rate per predator and β is the number of predators produced per prey killed. Having arrived at a specific model, Rosenzweig and Schaffer examine the dependence of the fitness of each species upon the various parameters through which genetic change must be expressed. Fitness is the true per capita growth rate ($m_H = 1/H \, dH/dt$, $m_P = 1/P \, dP/dt$). To simplify this task a bit, they note that the fitness of the prey is insensitive to either β or d and the fitness of the predator depends not at all on r or K. The possibilities for coevolution would appear to reside in the functional response parameters, a and h; Rosenzweig and Schaffer would have us determine the stability of the evolving interaction by looking at the influence of a and h upon the position of the predator isocline, arguing that evolution in the predatory species will tend to force the isocline to the left and evolution in the prey will force it to the right. The question is whether or not a coevolutionary steady state results, i.e. is $\hat{P}$ stable and to the right of the hump in the prey isocline? I shall not follow the argument in detail except to note that they arrived at simple, reasonable, and sufficient conditions for dynamical stability which in some circumstances also ensure a coevolutionary steady state. The existence of evolutionarily stable equilibria, therefore, does not appear unreasonable.

A few problems remain with this theory before we can allow it to lull us into the assurance that predator–prey systems will naturally evolve to stable equilibria. Built into this approach is some inevitable uncertainty about the structural stability of the result, i.e. would the same conclusions emerge if different forms of the functions g, f, and k had been used? In this case I suspect it would.

But the most critical question relates to the assumption that a mutation will affect only one parameter, independently of the others. The assignment of a mutation in the population to just one of a number of growth and interaction parameters is bound to be to some extent unjustified. Normally, for the sake of parsimony and in the absence of information to the contrary, we could justify it, but in this case both experimental evidence and a long tradition of natural history suggest that antipredator and life history evolution are correlated. Zareh et al. (1980) exposed houseflies to *Nasonia* for 25 generations and then compared the selected strain to control stocks. Houseflies did indeed rapidly evolve resistance; pupal weights increased, perhaps involving a strengthening of the puparium, and the pupal period shortened. But the price for this was a substantial decline in the biotic potential of the resistant strain. The mean number of adults to successfully emerge from eggs laid by females surviving 5 or more days after emergence numbered 666 in the control group and only 478 in the experimental, a difference probably due to a combination of decreased pupal survival and decreased fecundity.

What does this do to Rosenzweig's and Schaffer's argument? Zareh et al. suggest that the evolution of resistance will not only shift the predator isocline to the right but will also scale down or compress the prey isocline. In other words, this sort of correlation may buttress arguments for a co-evolutionary steady state. Perhaps they are correct, but if resistance is not negatively correlated with K but with r or with some combination of both, then it is not clear what will happen. And whether correlations among other parameters will have the same effect has yet to be determined.

Rosenzweig tried to address this problem in his 1973 paper (1973b) with a general argument dividing all selective aspects of the environment into consonant functions and dissonant functions. Consonant functions tend to increase the selective value of a mutation which influences equilibrial densities in a beneficial way; dissonant functions do the opposite. He did not discuss the relationship between consonant and dissonant functions and the primary beneficial mutations. I anticipate that the nature of such relationships will need to be clearly understood before his logic bears much fruit. But the argument serves at least to highlight the problem.

Both theoretical approaches share one limitation: they apply only to point equilibria and their stability. The next stage in the development of a theory of the evolution of predator–prey systems should be to build the evolutionary process into a model in such a way that it is more flexible than Levin's approach but has a more explicit role in dynamics than Rosenzweig's theory will allow. Since the process is one in which two antagonists each try to maximize their respective gain, it should be vulnerable to a game theoretic analysis. Stewart-Oaten (1982) has made a start in this direction, but the theory of differential or continuous games will probably be necessary for much progress (Isaacs, 1965). Such an approach could reveal more complicated evolutionary influences upon predatory dynamics than has heretofore been suspected, especially in that region of the phase-plane exhibiting limit cycles.

What sort of experimental evidence exists to reflect upon the idea of a coevolutionary steady state? Paynter and Bungay (1969) grew the bacterium *Escherichia coli* with its viral parasite, the T_2 bacteriophage, in a chemostat culture. Their experiments (Fig. 11.2) show an increase in phage-resistant mutant forms of the bacterium over time. Interestingly enough, sensitive cells did not disappear from the culture; apparently the greater growth rate of such cells partially compensated for their vulnerability.

In a similar set of chemostat experiments with T_3 and T_4 phage, Horne (1970) found cultures to exhibit typically a period of drastic fluctuation in abundance followed by a period of relative stasis in which baterial densities rose to the limits of the medium (Fig. 11.3). Both the bacteria and phage were found to be genetically different by virtue of their coexistence; the bacteria evolved resistance to phage attack and the phage became less virulent, exhibiting an extended latent period within its host cell.

Lin Chao *et al.* (1977) provide much more detail on an *E. coli*–T_7 phage system but with substantially the same result. Figure 11.4(a) shows one of their cultures in which *E. coli* did not develop resistance, and Fig. 11.4(b) reveals the effects of a resistant mutant arising in the bacterial population.

One might complain that parasitic relationships such as these are not subject to the same sort of evolutionary pressures as are predator–prey

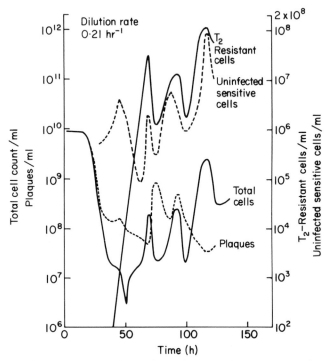

Figure 11.2 Chemostat populations of *E. coli* and T₂ bacteriophage (from Paynter and Bungay, 1969).

systems. I am not sure that this is a valid objection for an *E. coli*–bacteriophage system in which the host cell is destroyed upon successful completion of the parasite's development. But the complaint is most easily deflected by the observation that the same thing can occur in more traditional microbial predator–prey systems. Drake (1975) grew *E. coli* as prey for the vegetative, amoeboid stage of the slime mold *Dictyostelium discoideum*, in chemostats. He found that after 3–4 weeks of continuous culture the characteristic oscillations of this system ceased; the bacteria increased in density and were found to be resistant to predation.

Some caution is called for in automatically inferring coevolution from changes in dynamic properties of a culture. The study by Tsuchiya *et al.* (1972) of the interaction between *E. coli* and *Dictyostelium discoideum* (referred to in Chapter 5) gave evidence of stabilization after 3–4 weeks. But this developed from a physiological rather than a genetic change. Replicate cultures could be depended upon to stabilize at the same time.

And then, of course, there is the classic set of experiments done by Pimentel's laboratory on the housefly and its pupal parasitoid, *Nasonia vitripennis*. In the course of trying to get the two species to coexist (Chapter 7), Pimentel and his students noticed a qualitative change in dynamics which was later discovered

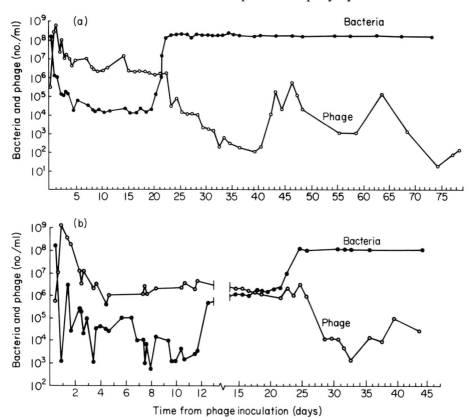

Figure 11.3 Population dynamics of *E. coli* and T_3 bacteriophage in replicate chemostat cultures (from Horne, 1970).

to result from a genetic change in the housefly's resistance to parasitism. For example, Fig. 11.5 from Pimentel and Stone (1968) displays the severe fluctuation of a newly constituted parasite–host culture and the much reduced oscillations of an older culture in which coevolution has occurred. Little doubt exists that these differences reflect genetic change; strain differences sufficient to produce them have been well documented (e.g. Chabora, 1972; Zareh *et al.*, 1980).

Since this result indicates such striking stabilization, we might reasonably ask if it is representative of host–parasitoid interactions in general. The answer is not clear. Chabora and Pimentel (1970) attempted to document similar coevolutionary stabilization of a culture in which *N. vitripennis* parasitized the blowfly, *Phaenicia sericata*; they had not nearly such dramatic results as had been obtained with houseflies. To the rejoinder that perhaps the *Nasonia*-blowfly interaction might not be as close to a coevolutionary equilibrium at the outset as the *Nasonia*-housefly interaction, one can only

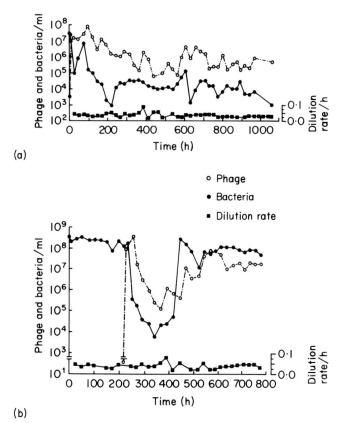

Figure 11.4 Chemostat cultures of *E. coli* and T$_7$ bacteriophage in which (a) a resistant phenotype of *E. coli* did not develop and (b) a resistant phenotype did develop (from Lin Chao *et al.*, 1977).

observe that while *Nasonia* is a general parasitoid of muscoid flies, blowflies are common hosts in the field and houseflies are not (Chabora, 1972).

Pimentel maintained that a comparable coevolutionary change may have taken place in Utida's classic study of the Azuki bean weevil, *Callosobruchus chinensis*, and its parasitoid, *Anisopteromalus calandrae*[7] (Utida, 1957). Utida maintained the two species in a simple culture (beans in a petri dish) for over 6 years (Fig. 11.6)! Examination of this unmatched time series suggests that the stability properties changed from about the 40th to the 90th generations. But the maintenance of a culture for 6 years with no variation in either the physical environment or the quality of the food supply for the host is virtually impossible. Simple differences in temperature can change the stability properties of host–parasite systems (Burnett, 1958). To suggest that the change in stability resulted from genetic alteration of the host or parasite without documentation of strain differences is unjustified.

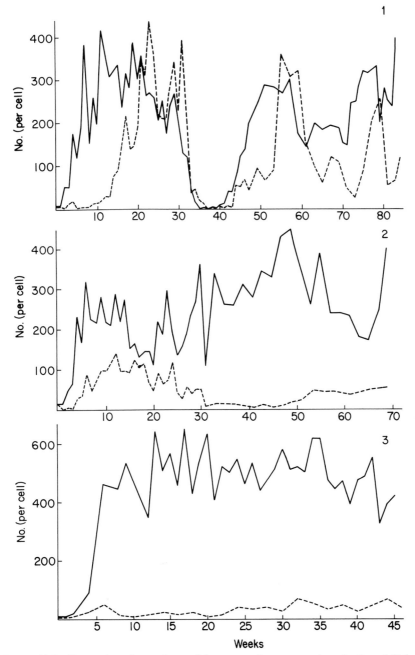

Figure 11.5 Dynamics of experimental host–parasite systems (graphs 2 and 3) by comparison to a control system (graph 1). Host numbers are solid lines; wasps are dashed lines. Houseflies in the control culture were of wild type; those in the experimental came from stocks selected for resistance to parasitism (from Pimentel and Stone, 1968).

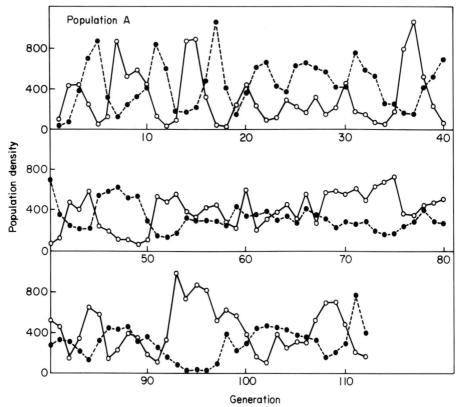

Figure 11.6   Population fluctuations of *C. chinensis* (solid lines and open circles) and its parasitoid *A. calandrae* (dashed lines and solid circles) (from Utida, 1957).

The only fair conclusion to draw from this body of experimental work is that natural selection can promote stability. However we must bear in mind that we have data for very few predator–prey systems and these only in the laboratory. It should surprise nobody if, in a few years' time, some enterprising doctoral student comes up with a counter-example.

NOTES

1. Quoted in Egerton (1973).
2. The two positions are, in the words of their advocates;

   When an ecologist investigates a species he may ask: given the existing characteristics of the species ... what determines the distribution and numbers of species in the world? In order to answer that question we do not need to know how the species evolved its particular characteristics. The phylogenetic question is interesting in itself but it is not relevant to the investigation of the question of distribution and abundance. How the species acquired its present adaptive characteristics is a second and independent question. (*Birch and Ehrlich (1967)*.)

In a herbivore-plant system, animal density influences selective pressure on plants; this selection influences genetic make-up of plants; and in turn, the genetic make-up of plants influences animal density. The actions and reactions of interacting populations in the food chain cycling in the genetic feed-back mechanism result in the evolution and regulation of animal populations. (*Pimentel (1961)*.)

3. Murie (1944), Rudebeck (1950, 1951) and Mech (1970); there are, however, exceptions: Hornocker (1970), Maher (1970) and Nellis *et al.* (1972). Selectivity of predators seems to be greatest when they are about the same size or smaller than their prey and capture is preceded by pursuit. Such an encounter is more likely to reveal structural or physiological defects than an ambush style of attack. Reports of non-selective predation usually deal with ambushing or a large size difference between prey and predator.

4. Slobodkin stated it this way:

    it is possible to solve for the population efficiency that would be associated with a predation procedure that removed only one size or age category of yield animals. ... The value of the concept of population efficiency ... lies in the fact that it relates yield production to population size and therefore permits us to outline the most efficient procedure for a predator. A prudent predator should take yield organisms in such a way as to maximize its yield and at the same time maximize the population efficiency of the prey. (1961, p. 142).

5. Michod (1979) supports this contention, although by different logic.

6. A sample: Pimentel (1961, 1968); Pimentel and Al-Hafidh (1965); Pimentel *et al.* (1963); Pimentel and Stone (1968); Chabora and Pimentel (1970); Chabora (1972) and Zareh *et al.* (1980).

7. Originally identified by Utida as *Neocatalaccus mamezophagus*.

12 Predation and the ecological community

> Where one animal preys on others of almost
> every class, ... the profusion of any one of
> these last may cause all such general feeders to
> subsist more exclusively upon the species thus
> in excess, and the balance may thus be
> restored.
>
> *Lyell (1832)*

The major purpose of this book has been to explore what happens when the boundaries of one's ecological system are expanded from consideration of only a single species to include both a predator and its prey. The lesson I have tried to communicate is that the class of potential population behaviors is so enlarged by such a broadening of perspective that the traditional concepts of single-species dynamics no longer provide a reliable theoretical guide. The question now remaining is what happens when the bounds of the system are drawn even more broadly?

Evidence seems to be accumulating that trophic interactions can be fundamental to the structure and operation of natural communities. Shallow-water marine communities exhibit complicated food-web structures, and predation appears to be critical to the maintenance of that structure.[1] And since the publication of the seminal paper by Brooks and Dodson in 1965, limnologists have been arguing about the importance of predation both to the structuring of fish and zooplankton communities and to the productivity and nutrient dynamics of lakes.[2] These are interesting and important areas of research in contemporary ecology, but I choose not to treat them in this book.[3] I seek in this chapter merely to inquire into the extent to which an ecologist must consider more than two species when searching for an explanation of predator–prey dynamics.

In such simple systems as Utida's Azuki bean beetles in the laboratory (Chapter 11), nothing beyond the beetle and its parasite are important. The beans themselves can be treated simply as an upper bound on the size of the beetle population. But is this a realistic model for the natural world? The wolves and moose of Isle Royale (Chapter 4) comprise only a slightly more complicated system, at first glance. Moose have no other predators or competitors; wolves have no other prey, except beaver which are only occasionally an important food source. The only other elements of the biotic environment important to this system are the plants which provide browse for the moose. But these plants are unlike beans in a petri dish in two critical ways: (1) their destruction by high densities of moose induces a time lag before recovery of biomass, and (2) the composition of the recovered plant community is different, being comprised of less vulnerable and nutritious species (P. Jordan, personal communication). In this circumstance the vegetation probably ought to enter into the conceptual scheme at a more fundamental level than as merely imposing an upper limit on the number of moose.

Perhaps the appropriate point to begin considering the effects of additional species are simple three-species systems, several of which merit consideration. Here and in Chapter 10 I have mentioned one kind of three-species system, that of a predator, a herbivore and food plants. Such three-level systems exist in other forms, such as a herbivore, its parasitoid and a hyperparasitoid,[4] but the carnivore–herbivore–plant food chain is probably the more common. Other classes of three-species systems are two predators consuming one prey species and one predatory species consuming two prey.

The two predator–one prey system might be considered, initially, the easier to characterize because it comprises a competition system, about which we know a great deal. Given the competitive exclusion principle, we might expect *a priori* that such systems would be unstable. Not much existing data bears on this topic. The little we have lends support to such an expectation, with one spectacular exception. First, the support: Fujii (1983) established a number of cultures of the Mexican bean beetle with two of its parasitic wasps, *Anisopteromalus calandrae* and *Heterospilus prosopidis*, to discover that none of these three-species cultures persisted long and in no case was an unstable two-species system stabilized by the addition of a second parasitoid. Then, the exception: Utida conducted a well-known set of experiments using the same parasitoids attacking the Azuki bean weevil (reported in Utida, 1957). He was able to keep the three species in culture for about four years (Fig. 12.1). This is a remarkable result, although one should bear in mind that only one of several replicates so persisted (Fujii, 1983).

One could explain away the persistence of this culture by an *ad hoc* subdivision of the Azuki bean weevil prey into independent 'resources' and so preserve the concept of competitive exclusion. Armstrong and McGehee (1980) offer an alternative explanation: they point out that in theory many different predators can coexist on only a few prey if those prey oscillate in abundance. The idea is that niche partitioning among the predators occurs in such a way as to make them specialists upon a small range of prey densities. For example, when prey are sparse and take refuge in a relatively inaccessible habitat, one predatory species might evolve a specialized phenotype which

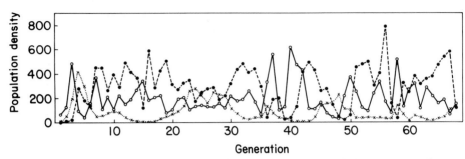

Figure 12.1 Population dynamics of the azuki bean weevil (solid lines, open circles) with two parasitoids, *A. calandrae* (dashed lines, solid circles), and *H. prosopidis* (dotted lines, x's) (from Utida, 1957).

allows it to exploit prey in the protected habitat; another might adopt a strategy of superior searching ability, so that as soon as prey numbers exceed the capacity of the refuge it could rapidly find the unprotected individuals; a third predator might opt for an ambush style of hunting most suited to high prey densities and develop a slow metabolism to make most efficient use of the prey it captures, and so on. Were prey numbers at a fixed equilibrium density, one of these predators would be superior for sufficiently long to outcompete the others. But a prey population that fluctuates may allow them to coexist, and both theory and the real world give us ample reason to believe in the ubiquity of fluctuation in predator–prey systems. If oscillations were sufficiently predictable this density niche could be subdivided more finely; in fact Armstrong and McGehee cite Zicarelli (1975) as having demonstrated the theoretical possibility of an infinite number of predators coexisting upon a single fluctuating prey species.

Utida's results seem to be best explained by such a mechanism. In his 1957 paper he observed that the relative advantage of the two wasps reversed as a function of host density (Fig. 12.2). One was a good searcher; the other had a longer ovipositor to reach well-hidden larvae. Were host densities constant, he suggested, one or the other species of wasp would have gone extinct; but given a constantly-changing beetle density, neither wasp eliminated the other.

Since oscillations do not occur in simple models of competition, the concepts which have emerged from those models, such as competitive exclusion, may not be strictly applicable to the coexistence of competing predators. Perhaps the conservative conclusion would be that the stability of two predator–one prey systems will probably not exceed that of the more stable one predator–one prey subsystem.

Substantially more attention has been given to the third case, that of a single predator attacking two species of prey, prey which may be either ecologically unrelated or competitors. In either case a powerful stabilizing mechanism is available in the switching functional response (Chapter 8). When a predator can avoid starvation in the face of a decrease in the numbers of one species by

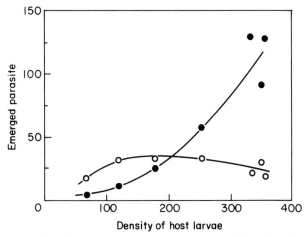

Figure 12.2   Number of parasites to emerge from a range of densities of host larvae for the two wasps *A. calandrae* (solid circles) and *H. prosopidis* (open circles), when both attack simultaneously (from Utida, 1957).

switching to another, the original prey may both be spared excessive mortality when its density is low and be prevented from a subsequent explosive increase following the starvation of its predators. For example, McMurtry and Scriven (1966) found that the predatory mite, *Amblyseius hibisci*, coexisted stably and at low densities with its prey, *Oligonychus punicae*, if pollen was provided as an alternate food source. When pollen was not provided, the herbivorous mite periodically exploded in numbers to the limits of its food supply.

The more popular circumstance occurs when prey species compete with each other. This class of predator–prey systems has received a great deal of attention, primarily because of the possibility that predation may serve to mitigate the competitive interaction and allow a potentially unstable association to persist.[5] Less attention has been given to the influence of the addition of an alternative competing prey species upon the actual dynamics of the predator–prey interaction. Certainly switching is possible, and we might expect competitors to vary in numbers inversely, always providing the predator with a prey base. On logical grounds it would not be unreasonable to expect such systems to be stable (Inouye, 1980). Given that nearly all ecologists believe this result, it is a bit surprising that almost no empirical evidence exists to confirm it. I suspect that this reflects simple neglect, for the most part. We sometimes consider a concept so reasonable that real-world tests are deemed unnecessary. Utida (1953) was able to stabilize competition between two bean beetles with the introduction of a parasite, but the only other attempt of which I am aware failed. Cornell and Pimentel (1978) observed *Nasonia vitripennis* to switch when it was given houseflies and the two blowflies, *Phaenicia sericata* and *Phormia regina*, but they saw no evidence that *Nasonia* stabilized competition.

With some uncertainty remaining about how three-species systems will operate, one truth emerges: their dynamics are bound to be enormously complicated by comparison to two-species systems, the dynamics of which are complicated enough. Intricate cyclic and chaotic behaviors atypical of two-species interactions can easily arise with the addition of a third species (Gilpin, 1979).

Relatively few efforts have been made to work with models of more than three species. Armstrong (1982) examined a general two-predator–two-prey system in which the predators can be made to vary (equally) between complete specialists on only one prey species and generalists with equal preference for either type. He demonstrated an interesting relationship between the stability of this interaction and the degree of preference the predators exhibit. As preference for one prey type becomes less pronounced, the two systems grow more closely coupled. By solving for all four 'modes of vibration' of the model near equilibrium, Armstrong was able to demonstrate that closer coupling simultaneously enhanced the stability of the prey species and decreased the stability of the predators. Since a system is only as stable as its least stable components, overall stability declined. This result was achieved by deliberately avoiding consideration of the full range of dynamical behaviors such a model could exhibit. Consequently Armstrong's effort should be considered as just the first stage in the exploration of the model.

The question we must eventually address is what limits on the complexity of a community must we impose if we are to gain insight into the dynamics of the constituent species' populations. The goals of theoretical ecologists would appear to place that limit at 3–4 species at present. Applied ecologists face much less severe constraints. Their ability to discover reasonable values for some parameters reduces drastically the difficulty of theoretical or numerical analysis. Perhaps an upper bound of six to seven species might be reasonable for such workers.

But the real world frequently exceeds even this limit. Figure 12.3 shows the benthic community of McMurdo Sound as described by Dayton *et al.* (1974). Approximately 55% of the bottom substratum is covered with sponges of 21 species. All of these grow slowly except for one, *Mycale acerata*, which grows fast and is potentially dominant over the others. But competition for space is uncommon because of predation by a complex of five asteroids and a nudibranch. *Mycale* in particular is suppressed by two asteroids, *Acodontaster conspicuus*, a large general sponge predator, and *Perknaster fuscus antarcticus*, a specialist on *Mycale*. Together these predators prevent *Mycale* from overwhelming the rest of the sponge community. Since *Acodontaster* is capable of killing even the largest sponge and since the standing crop of its prey is so large, one wonders why it does not explode in numbers along with another large-sponge predator, the dorid nudibranch, *Austrodoris mcmurdensis*. Two higher-level predators apparently prevent this. The small asteroid, *Odontaster validus*, both consumes the larvae of *Acodontaster* and *Austrodoris*

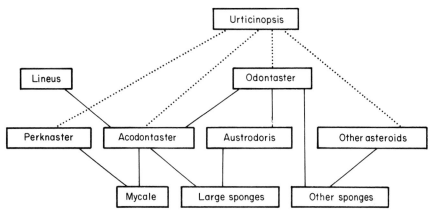

Figure 12.3   Food web of the benthic community of McMurdo Sound. The dotted lines are meant to imply only a weak or occasional predatory interaction between *Urticinopsis* and the asteroids.

as it feeds on detritus and also attacks adult *Acodontaster*. *Odontaster* digests a hole in the aboral surface of *Acodontaster* and is joined in the attack by the large nemertean, *Lineus corrugatus*. A large sessile anemone, *Urticinopsis antarctica*, will also consume *Acodontaster*. Together these secondary predators appear to be enough to forestall an increase in the sponge predators.

One could, in principle, isolate any predator–prey pair from this community for consideration alone but the likelihood of saying anything meaningful about a system so defined is small. Of course the likelihood of saying anything meaningful about the dynamics of the community *in toto* is even smaller.

Efforts to extend concepts from simple theory to such large communities will almost certainly run into both practical and theoretical problems. Practical problems have to do with the design and evaluation of experiments where sometimes hundreds of different manipulative treatments are available. Theoretical problems relate to the unfortunate fact that large models with many equations may be sufficiently difficult to understand that doing experiments with the real community may be as easy as trying to understand the model. At this point it looks as though students of whole communities will need to be satisfied with a descriptive approach to static evolutionary patterns.

NOTES

1. Probably the bulk of the experimental work on predation's role in the shaping of community structure has been conducted in the intertidal and shallow subtidal zone. A review of earlier work may be found in Connell (1972) and of more recent work in Paine (1980) and Dayton (1984).
2. Zaret (1980) provides a recent review of this literature.

3. The empirical and theoretical literature on food webs is reviewed by Pimm (1982), an earlier volume in this series.
4. Beddington and Hammond (1977) suggest that the constraints upon such three-level systems for stability may be more severe than upon simple host–parasitoid systems.
5. A variety of theory gives support to the idea (Caswell, 1978; Vance, 1978; Inouye, 1980). So does some experimental work (Slobodkin, 1964).

Appendix

Common name	Scientific name
Azuki bean weevil	*Callosobruchus chinensis*
Badger	*Taxidea taxus*
Beaver	*Castor canadensis*
Black vulture	*Coragyps atratus*
Bobwhite quail	*Colinus virginianus*
Caribou	*Rangifer tarandus*
Carrion crow	*Corvus corone*
Cod	*Gadus morhua*
Cotton rat	*Sigmodon hispidus*
Cottontail rabbit	*Sylvilagus* spp.
Cottony-cushion scale	*Icerya purchasi*
Coyote	*Canis latrans*
Cyclamen mite	*Tarsonemus pallidus*
Dall sheep	*Ovis dalli*
Deer mouse	*Peromyscus leucopus*
Elk	*Cervus elaphus*
European pine sawfly	*Neodiprion sertifer*
Fisher	*Martes pennanti*
Forster's tern	*Sterna forsteri*
Fox	*Vulpes vulpes*
Frigate bird	*Fregata magnificens*
Garter snake	*Thamnophis sirtalis*
Ghost crab	*Ocypode occidentalis*
Goshawk	*Accipiter nisus*
Grasshopper mouse	*Onychomys torridus*
Great horned owl	*Bubo virginianus*
Great-tailed grackle	*Cassidix mexicanus*
Herring	*Clupea harengus*
Housefly	*Musca domestica*
Jackrabbit	*Lepus townsendii*
Lime aphid	*Eucallipterus tiliae*
Lion	*Felis leo*
Lynx	*Felis lynx*
Mexican bean beetle	*Zabrotes subfaciatus*
Mink	*Mustela vison*
Mountain lion	*Felis concolor*

Common name	Scientific name
Moose	*Alces alces*
Mule deer	*Odocoileus hemionus*
Muskrat	*Ondatra zibethicus*
Olive scale	*Parlatoria oleae*
Oystercatcher	*Haematopus ostralegus*
Pacific Ridley turtle	*Lepidochelys olivacea*
Partridge	*Perdix perdix*
Periodical cicada	*Magicicada* spp.
Pheasant	*Phasianus colchicus*
Raccoon	*Procyon lotor*
Ruffed grouse	*Bonasa umbellus*
Saber-toothed cat	*Smilodon californicus*
Saithe	*Pollachius virens*
Snowshoe hare	*Lepus americanus*
Spotted hyena	*Crocata crocata*
Spotted skunk	*Spilogale putorius*
Spruce budworm	*Archips fumiferana*
Stickleback	*Gasterosteus aculeatus*
Stoat	*Mustela erminea*
Striped skunk	*Mephitis mephitis*
Titmouse	*Parus* spp.
Tsetse fly	*Glossina* spp.
Uinta ground squirrel	*Spermophilus armatus*
Vedalia	*Rodolia cardinalis*
White-winged dove	*Zenaida asiatica*
Wildebeest	*Connochaetes taurinus*
Wolf	*Canus lupus*

Bibliography

Abrams, P. A. (1982), Functional responses of optimal foragers. *The American Naturalist*, **120**, 382–390.

Abramsky, Z. and Tracy, C. R. (1979), Population biology of a 'noncycling' population of prairie voles and a hypothesis on the role of migration in regulating microtine cycles. *Ecology*, **60**, 349–361.

Akre, B. G. and Johnson, D. M. (1979), Switching and sigmoid functional response curves by damselfly naiads with alternate prey available. *Journal of Animal Ecology*, **48**, 703–720.

Armstrong, R. A. (1976), The effects of predator functional response and prey productivity on predator–prey stability: a graphical approach. *Ecology*, **57**, 609–611.

Armstrong, R. A. (1982), The effects of connectivity on community stability. *American Naturalist*, **120**, 391–402.

Armstrong, R. A. and McGehee, R. (1980), Competitive exclusion. *The American Naturalist*, **115**, 151–170.

Auslander, D. M., Oster, G. F. and Huffaker, C. B. (1974), Dynamics of interacting populations. *Journal of the Franklin Institute*, **297**, 345–376.

Ball, I. R. (1975), Nature and formulation of biogeographical hypotheses. *Systematic Zoology*, **24**, 407–430.

Balser, D. S., Dill, H. H. and Nelson, H. K. (1968), Effect of predator reduction on waterfowl nesting success. *Journal of Wildlife Management*, **32**, 669–682.

Banks, C. J. (1959), The behaviour of individual coccinellid larvae on plants. *British Journal of Animal Behaviour*, **5**, 12–24.

Bänsch, R. (1966), On prey-seeking behaviour of aphidophagous insects, in *Ecology of aphidophagous insects* (ed. I. Hodeck), pp. 123–128, Academia, Prague.

Baumer, F. L. V. (1977), *Modern European thought: continuity and change in ideas, 1600–1950*. Macmillan, New York.

Beddington, J. R. and Free, C. A. (1976), Age structure effects in predator–prey interactions. *Theoretical Population Biology*, **9**, 15–24.

Beddington, J. R. and Hammond, P. S. (1977), On the dynamics of host–parasite–hyperparasite interactions. *Journal of Animal Ecology*, **46**, 811–822.

Beddington, J. R., Free, C. A. and Lawton, J. H. (1978), Characteristics of successful natural enemies in models of biological control of insect pests. *Nature*, **273**, 513–519.

Bergerud, A. T. (1971), The population dynamics of Newfoundland caribou. *Wildlife Monographs*, **25**, 1–55.

Beukema, J. J. (1968), Predation by the three-spined stickleback (*Gasterosteus aculeatus* L.): the influence of hunger and experience. *Behaviour*, **31**, 1–126.

Birch, L. C. and Ehrlich, P. R. (1967), Evolutionary history and population biology. *Nature*, **214**, 349–352.

Birney, E. C., Grant, W. E. and Baird, D. D. (1976), Importance of vegetative cover to cycles of *Microtus* populations. *Ecology*, **57**, 1043–1051.

Blankinship, D. R. (1966), The relationship of white-winged dove production to control of great-tailed grackles in the lower Rio Grande Valley of Texas. *Transactions of the North American Wildlife and Natural Resources Conference*, **31**, 45–58.

Brock, V. E. and Riffenburgh, R. H. (1960), Fish schooling: a possible factor reducing predation. *Journal du Conseil international pour l'Exploration de la Mer*, **25**, 305–317.

Brooks, J. L. and Dodson, S. I. (1965), Predation, body size, and composition of plankton. *Science*, **150**, 28–35.

Buckley, F. G. and Buckley, P. A. (1974), Comparative feeding ecology of wintering adult and juvenile royal terns (Aves: Laridae, Sterninae). *Ecology*, **55**, 1053–1063.

Bulmer, M. G. (1974), A statistical analysis of the 10-year cycle in Canada. *Journal of Animal Ecology*, **43**, 701–718.

Burnett, T. (1958), Effect of host distribution on the reproduction of *Encarsia formosa* Gahan (Hymenoptera: Chalcidoidea). *Canadian Entomologist*, **90**, 179–191.

Burnett, T. (1964). An acarine predator–prey population infesting stored products. *Canadian Journal of Zoology*, **42**, 655–673.

Caltagirone, L. E. (1981), Landmark examples in classical biological control. *Annual Review of Entomology*, **26**, 213–232.

Canale, R. P. (1970), An analysis of models describing predator–prey interactions. *Biotechnology and Bioengineering*, **12**, 353–378.

Caswell, H. (1978), Predator-mediated coexistence: a nonequilibrium model. *The American Naturalist*, **112**, 127–154.

Chabora, P. C. (1972), Studies in parasite-host interaction. IV. Modification of parasite, *Nasonia vitripennis*, responses to control and selected host, *Musca domestica*, populations. *Annals of the Entomological Society of America*, **65**, 323–328.

Chabora, P. C. and Pimentel, D. (1970), Patterns of evolution in parasite–host systems. *Annals of the Entomological Society of America*, **63**, 479–486.

Charlesworth, B. (1980), *Evolution in Age-structured Populations*, Cambridge University Press, Cambridge.

Chesness, R. A., Nelson, M. M. and Longley, W. H. (1968), The effect of predator removal on pheasant reproductive success. *Journal of Wildlife Management*, **32**, 683–697.

Chesson, J. (1978), Measuring preference in selective predation. *Ecology*, **59**, 211–215.

Chesson, J. (1983), The estimation and analysis of preference and its relationship to foraging models. *Ecology*, **64**, 1297–1304.

Chitty, D. (1960), Population processes in the vole and their relevance to general theory. *Canadian Journal of Zoology*, **38**, 99–113.

Chitty, D. (1967a), The natural selection of self-regulating behaviour in animal populations. *Proceedings of the Ecological Society of Australia*, **2**, 51–78.

Chitty, D. (1967b), What regulates bird populations? *Ecology*, **48**, 698–701.

Clark, L. R., Geier, P. W., Hughes, R. D. and Morris, R. F. (1967), *The Ecology of Insect Populations in Theory and Practice*, Methuen and Co., London.

Clausen, C. P. (ed.) (1977), *Introduced Parasites and Predators of Arthropod Pests and Weeds: a World Review*. United States Department of Agriculture, Agriculture Handbook Number 480.

Connell, J. H. (1970), A predator–prey system in the marine intertidal region. I. *Balanus glandula* and several predatory species of *Thais*. *Ecological Monographs*, **40**, 49–78.

Connell, J. H. (1972), Community interactions on marine rocky intertidal shores. *Annual Review of Ecology and Systematics*, **3**, 169–192.

Cornell, H. and Pimentel, D. (1978), Switching in the parasitoid *Nasonia vitripennis* and its effect on host competition. *Ecology*, **59**, 297–308.

Craighead, F. C., Jr. and Craighead, J. J. (1956), *Hawks, owls, and wildlife*, Stackpole Co., Harrisburg, and Wildlife Management Institute, Washington.

Crête, M., Taylor, R. J. and Jordan, P. A. (1981a), Simulating conditions for the regulation of a moose population by wolves. *Ecological Modelling*, **12**, 245–252.

Crête, M., Taylor, R. J. and Jordan, P. A. (1981b), Optimization of moose harvest in southwestern Quebec. *Journal of Wildlife Management*, **45**, 598–611.

Crook, J. H. (1964), The evolution of social organization and visual communication in the weaver birds (*Ploceinae*). *Behavior*, Supplement **10**, 1–178.

Crook, J. H. (1965), The adaptive significance of avian social organizations. *Symposium of the Zoological Society of London*, **14**, 181–218.

Croze, H. (1970), Searching image in carrion crows. *Zeitschrift fur Tierpsychologie*, Supplement 5, 1–86.

Curio, E. (1976), *The Ethology of Predation*, Springer Verlag, Berlin.

Curry, G. L. and DeMichele, D. W. (1977), Stochastic analysis for the description and synthesis of predator–prey systems. *Canadian Entomologist*, **109**, 1167–1174.

Darwin, C. (1859), *On the origin of species by means of natural selection, or the preservation of favoured races in the struggle for life*, John Murray, London.

Dayton, P. K. (1984), Processes structuring some marine communities: are they general? in *Ecological Communities: Conceptual Issues and the Evidence*. (eds. L. G. Abele, D. S. Simberloff, D. R. Strong and A. Thistle), Princeton University Press, Princeton.

Dayton, P. K., Robilliard, G. A., Paine, R. T. and Dayton, L. B. (1974), Biological accommodation in the benthic community at McMurdo Sound, Antarctica. *Ecological Monographs*, **44**, 105–128.

Dixon, A. F. G. (1959), An experimental study of the searching behavior of the predatory coccinellid beetle *Adalia decempunctata* (L.). *Journal of Animal Ecology*, **28**, 259–281.

Dixon, A. F. G. (1971), The role of intra-specific mechanisms and predation in regulating the numbers of the lime aphid, *Eucallipterus tiliae* L. *Oecologia*, **8**, 179–193.

Drake, J. F. (1975), *Effect of environment on microbial prey–predator interactions*. Ph.D. Dissertation, University of Minnesota.

Duebbert, H. F. and Kantrud, H. A. (1974), Upland duck nesting related to land use and predator reduction. *Journal of Wildlife Management*, **38**, 257–265.

Edminster, F. C. (1939), The effect of predator control on ruffed grouse populations in New York. *Journal of Wildlife Management*, **3**, 345–352.

Egerton, F. N. (1973), Changing concepts of the balance of nature. *Quarterly Review of Biology*, **48**, 322–350.

Elseth, G. D. and Baumgardner, K. D. (1981), *Population biology*, D. Van Nostrand Co., New York.

Elton, C. S. (1924), Periodic fluctuations in the number of animals: their causes and effects. *British Journal of Experimental Biology*, **2**, 119–163.

Elton, C. S. (1927), *Animal ecology*, Sidgwick and Jackson, London.

Elton, C. S. and Nicholson, M. (1942), The ten-year cycle in numbers of the lynx in Canada. *Journal of Animal Ecology*, **11**, 215–244.

Emlen, J. M. (1966), The role of time and energy in food preference. *The American Naturalist*, **100**, 611–617.

Errington, P. L. (1946), Predation and vertebrate populations. *Quarterly Review of Biology*, **21**, 144–177; 221–245.

Errington, P. L. (1956), Factors limiting vertebrate populations. *Science*, **124**, 304–307.

Errington, P. L. (1963a), The phenomenon of predation. *American Scientist*, **51**, 180–192.

Errington, P. L. (1963b), *Muskrat populations*, Iowa State University Press, Ames.

Errington, P. L. (1967), *Of predation and life*, Iowa State University Press, Ames.

Finerty, J. P. (1980), *The population ecology of cycles in small mammals*, Yale University Press, New Haven.

Fisher, R. A. (1930), *The genetical theory of natural selection*, Oxford University Press, London.

Flanders, S. E. (1948), A host–parasite community to demonstrate balance. *Ecology*, **29**, 123.

Flanders, S. E. and Badgley, M. E. (1963), Prey–predator interactions in self-balanced laboratory populations. *Hilgardia*, **35**, 145–183.

Fleschner, C. A. (1950), Studies on search capacity of the larvae of three predators of the citrus red mite. *Hilgardia*, **20**, 233–265.

Formanowicz, D. R., Jr. (1982), Foraging tactics of larvae of *Dytiscus verticalis* (Coleoptera: Dytiscidae): the assessment of prey density. *Journal of Animal Ecology*, **51**, 757–768.

Fox, L. R. (1975), Cannibalism in natural populations. *Annual Review of Ecology and Systematics*, **6**, 87–106.

Freedman, H. I. (1980), *Deterministic mathematical models in population ecology*, M. Dekker, Inc., New York.

Fujii, K. (1983), Resource dependent stability in an experimental laboratory resource-herbivore-carnivore system. *Researches on Population Ecology*, Supplement 3, 15–26.

Galton, F. (1883), *Inquiries into Human Faculty and its Development*, Macmillan, London.

Garsd, A. and Howard, W. E. (1981), A 19-year study of microtine population fluctuations using time-series analysis. *Ecology*, **62**, 930–937.

Gatto, M. and Rinaldi, S. (1977), Stability analysis of predator–prey models via the Liapunov method. *Bulletin of Mathematical Biology*, **39**, 339–347.

Gause, G. F. (1934), *The Struggle for Existence*, The Williams and Wilkins Co., Baltimore.

Gause, G. F., Smaragdova, N. P. and Witt, A. A. (1936), Further studies of interaction between predators and prey. *Journal of Animal Ecology*, **5**, 1–18.

Gilpin, M. E. (1972), Enriched predator–prey systems: theoretical stability. *Science*, **177**, 902–904.

Gilpin, M. E. (1975), *Group Selection in Predator–Prey Communities*, Princeton University Press, Princeton.

Gilpin, M. E. (1979), Spiral chaos in a predator–prey model. *The American Naturalist*, **113**, 306–308.

Gurney, W. S. C. and Nisbet, R. M. (1975), The regulation of inhomogeneous populations. *Journal of Theoretical Biology*, **52**, 441–457.

Haber, G. C. (1977), *Socio-ecological dynamics of wolves and prey in a subarctic ecosystem*. Ph.D. Dissertation, University of British Columbia.

Haber, G. C., Walters, C. J. and Cowan, I. McT. (1976), *Stability properties of a wolf–ungulate system in Alaska and management implications*. R–5–R, Institute of Resource Ecology, University of British Columbia.

Hamilton, W. D. (1966), The moulding of senescence by natural selection. *Journal of Theoretical Biology*, **12**, 12–45.

Hamilton, W. D. (1971), Geometry for the selfish herd. *Journal of Theoretical Biology*, **31**, 295–311.

Hardin, G. (1959), *Nature and Man's Fate*, Rinehart and Co., New York.

Hassell, M. P. (1966), Evaluation of parasite or predator responses. *Journal of Animal Ecology*, **35**, 65–75.

Hassell, M. P. (1978), *The Dynamics of Arthropod Predation*, Princeton University Press, Princeton.

Hassell, M. P. and May, R. M. (1974), Aggregation of predators and insect parasites and its effect on stability. *Journal of Animal Ecology*, **43**, 567–594.

Hastings, A. (1977), Spatial heterogeneity and the stability of predator–prey systems. *Theoretical Population Biology*, **12**, 37–48.

Hastings, A. (1978), Global stability of two species systems. *Journal of Mathematical Biology*, **5**, 399–403.

Hastings, A. and Wollkind, D. (1982), Age structure in predator–prey systems. I. A general model and a specific example. *Theoretical Population Biology*, **21**, 44–56.

Hempel, C. G. (1966), *Philosophy of natural science*, Prentice Hall, Englewood Cliffs.

Hilborn, R. (1979), Some long term dynamics of predator–prey models with diffusion. *Ecological Modelling*, **6**, 23–30.

Hobson, E. S. (1978), Aggregating as a defense against predators in aquatic and terrestrial environments, in *Contrasts in Behavior* (eds. E. S. Reese and F. J. Lighter), J. Wiley and Sons, Chichester.

Holling, C. S. (1959a), The components of predation as revealed by a study of small-mammal predation of the European pine sawfly. *Canadian Entomologist*, **91**, 293–320.

Holling, C. S. (1959b), Some characteristics of simple types of predation and parasitism. *Canadian Entomologist*, **91**, 385–398.

Holling, C. S. (1961), Principles of insect predation. *Annual Review of Entomology*, **6**, 163–182.

Holling, C. S. (1963), An experimental component analysis of population processes. *Memoirs of the Entomological Society of Canada*, **32**, 1–86.

Holling, C. S. (1964), The analysis of complex population processes. *Canadian Entomologist*, **96**, 335–347.

Holling, C. S. (1965), The functional responses of predators to prey density and its role in mimicry and population regulation. *Memoirs of the Entomological Society of Canada*, **45**, 1–60.

Holling, C. S. (1966), The functional response of invertebrate predators to prey density. *Memoirs of the Entomological Society of Canada*, **48**, 1–86.

Horn, H. S. (1968), The adaptive significance of colonial nesting in the Brewer's blackbird (*Euphagus cyanocephalus*). *Ecology*, **49**, 682–694.

Horne, M. T. (1970), Coevolution of *E. coli* and bacteriophage in chemostat culture. *Science*, **168**, 992–993.

Hornocker, M. G. (1969), Winter territoriality in mountain lions. *Journal of Wildlife Management*, **33**, 457–464.

Hornocker, M. G. (1970), An analysis of mountain lion predation upon mule deer and elk in the Idaho Primitive Area. *Wildlife Monographs*, **21**, 1–39.

Hsu, S. B. (1978), On global stability of a predator–prey system. *Mathematical Biosciences*, **39**, 1–10.

Huffaker, C. B. (1958), Experimental studies on predation. II. dispersion factors and predator–prey oscillations. *Hilgardia*, **27**, 343–383.

Huffaker, C. B. (1970), The phenomenon of predation and its role in nature, in *Proceedings of the Advanced Study Institute on the Dynamics of Numbers of Populations* (eds. M. H. den Boer and G. R. Gradwell), Oosterbeek.

Huffaker, C. B. and Kennett, C. E. (1956), Experimental studies on predation: predation and cyclamen mite population on strawberries in California. *Hilgardia*, **26**, 191–222.

Huffaker, C. B., Shea, K. P. and Herman, S. G. (1963), Experimental studies on predation. Complex dispersion and levels of food in an acarine predator–prey interaction. *Hilgardia*, **34**, 305–330.

Huffaker, C. B., Messenger, P. S. and DeBach, P. (1971), The natural enemy component in natural control and the theory of biological control, in *Biological Control* (ed. C. B. Huffaker), Plenum Press, New York.

Huffaker, C. B. and Matsumoto, B. M. (1982), Group versus individual functional responses of *Venturia* (= *Nemeritis*) *canescens* (Grav.). *Researches on Population Ecology*, **24**, 250–269.

Hughes, D. A. and Richard, J. D. (1974), The nesting of the Pacific Ridley turtle *Lepidochelys olivacea* on Playa Nancite, Costa Rica. *Marine Biology*, **24**, 97–107.

Inouye, R. S. (1980), Stabilization of a predator–prey equilibrium by the addition of a second 'keystone' victim. *The American Naturalist*, **115**, 300–305.

Isaacs, R. (1965), *Differential Games*, John Wiley and Sons, New York.

Isaacs, R. (1979), On applied mathematics. *Journal of Optimization Theory and Applications*, **27**, 31–50.

Ivlev, V. S. (1961), *Experimental Ecology of the Feeding of Fishes*, Yale University Press, New Haven.

Jacobs, J. (1974), Quantitative measurement of food selection: a modification of the forage ratio and Ivlev's electivity index. *Oecologia*, **14**, 413–417.

Jeffries, C. (1974), Qualitative stability and digraphs in model ecosystems. *Ecology*, **55**, 1415–1419.

Jordan, P. A., Shelton, P. C. and Allen, D. L. (1967), Numbers, turnover, and social structure of the Isle Royale wolf population. *American Zoologist*, **7**, 233–252.

Jordan, P. A. and Wolfe, M. L. (1980), Aerial and pellet-count inventory of moose at Isle Royale, in *Proceedings of the Second Conference on Scientific Research in the National Parks* (ed. J. Gogue), pp. 363–393, Washington, DC. U.S. National Park Service Transactions and Proceedings, Ser. 5.

Jost, J. L., Drake, J. F., Fredrickson, A. G. and Tsuchiya, H. M. (1973a), Interactions of

Tetrahymena pyriformis, Escherichia coli, Azotobacter vinelandii, and glucose in a minimal medium. *Journal of Bacteriology,* **113,** 834–840.

Jost, J. L., Drake, J. F., Tsuchiya, H. M. and Fredrickson, A. G. (1973b), Microbial food chains and food webs. *Journal of Theoretical Biology,* **41,** 461–484.

Keith, L. B. (1963), *Wildlife's Ten-year Cycle,* University of Wisconsin Press, Madison.

Keith, L. B. (1974), Some features of population dynamics of mammals, in *XI International Congress of Game Biologists,* Sweden.

Keith, L. B. and Windberg, L. A. (1978), A demographic analysis of the snowshoe hare cycle. *Wildlife Monographs,* **58,** 4–70.

Kenward, R. E. (1978), Hawks and doves: attack success and selection in goshawk flights at woodpigeons. *Journal of Animal Ecology,* **47,** 449–460.

Keyfitz, N. (1968), *Introduction to the Mathematics of Population,* Addison-Wesley, Reading.

Kierstead, H. and Slobodkin, L. B. (1953), The size of water masses containing plankton blooms. *Journal of Marine Research,* **12,** 141–147.

Krebs, C. J. (1978a), *Ecology: The Experimental Analysis of Distribution and Abundance,* 2nd edn, Harper and Row, New York.

Krebs, C. J. (1978b), A review of the Chitty hypothesis of population regulation. *Canadian Journal of Zoology,* **56,** 2463–2480.

Krebs, C. J. and Myers, J. H. (1974), Population cycles in small mammals. *Advances in Ecological Research,* **8,** 268–400.

Krebs, J. R., MacRoberts, M. H. and Cullen, J. M. (1972), Flocking and feeding in the great tit: an experimental study. *Ibis,* **114,** 507–530.

Kruuk, H. (1970), Interactions between populations of spotted hyenas and their prey species, in *Animal populations in relation to their food resources* (ed. A. Watson), British Ecological Society Symposium, **10,** 359–374.

Kruuk, H. (1972), *The Spotted Hyena.* University of Chicago Press, Chicago.

Lack, D. (1954), Cyclic mortality. *Journal of Wildlife Management,* **18,** 25–37.

Lechowicz, M. J. (1982), The sampling characteristics of electivity indices. *Oecologia,* **52,** 22–30.

Levin, S. A. (1972), A mathematical analysis of the genetic feedback mechanism. *The American Naturalist,* **106,** 145–164.

Levin, S. A. (1977), A more functional response to predator–prey stability. *The American Naturalist,* **111,** 381–383.

Levin, S. A. and Udovic, J. D. (1977), A mathematical model of coevolving populations. *The American Naturalist,* **111,** 657–675.

Lidicker, W. Z., Jr. (1973), Regulation of numbers in an island population of the California vole, a problem in community dynamics. *Ecological Monographs,* **43,** 271–302.

Lin Chao, Levin, B. R. and Stewart, F. M. (1977), A complex community in a simple habitat: an experimental study with bacteria and phage. *Ecology,* **58,** 369–378.

Lin, J. and Kahn, P. B. (1977), Qualitative behavior of predator–prey communities. *Journal of Theoretical Biology,* **65,** 101–132.

Lloyd, M. and Dybas, H. S. (1966), The periodical cicada problem. I. population ecology. *Evolution,* **20,** 133–149.

Lloyd, M. and Dybas, H. S. (1966), The periodical cicada problem. II. evolution. *Evolution,* **20,** 466–505.

Lotka, A. J. (1925), *Elements of Physical Biology*, Williams and Wilkins, Baltimore.

Luckinbill, L. S. (1973), Coexistence in laboratory populations of *Paramecium aurelia* and its predator *Didinium nasutum*. *Ecology*, **54**, 1320–1327.

Luckinbill, L. S. (1974), The effects of space and enrichment on a predator–prey system. *Ecology*, **55**, 1142–1147.

Lyell, C. (1832), *Principles of geology, being an attempt to explain the former changes in the earth's surface by reference to causes now in operation*, Vol. 2, John Murray, London.

MacArthur, R. H. and Pianka, E. R. (1966), On optimal use of a patchy environment. *The American Naturalist*, **100**, 603–609.

McKinney, H. L. (1966), Alfred Russell Wallace and the discovery of natural selection. *Journal of the History of Medicine and Allied Sciences*, **21**, 333–357.

McMurtrie, R. (1978), Persistence and stability of single-species and prey–predator systems in spatially heterogeneous environments. *Mathematical Biosciences*, **39**, 11–51.

McMurtry, J. A. and Scriven, G. T. (1966), Studies of predator–prey interactions between *Amblyseius hibisci* and *Oligonychus punicae* under greenhouse conditions. *Annals of the Entomological Society of America*, **59**, 793–800.

Madden, J. L. and Pimentel, D. (1965), Density and spatial relationships between a wasp parasite and its housefly host. *Canadian Entomologist*, **97**, 1031–1037.

Maher, W. J. (1970), The pomarine jaeger as a brown lemming predator in northern Alaska. *Wilson Bulletin*, **82**, 130–157.

Maiorana, V. C. (1976), Reproductive value, prudent predators, and group selection. *The American Naturalist*, **110**, 486–489.

Maly, E. J. (1969), A laboratory study of the interaction between the predatory rotifer *Asplanchna* and *Paramecium*. *Ecology*, **50**, 59–73.

Maly, E. J. (1978), Stability of the interaction between *Didinium* and *Paramecium*: effects of dispersal and predator time lag. *Ecology*, **59**, 733–741.

Mann, R. H. K. (1982), The annual food consumption and prey preferences of pike (*Esox lucius*) in the River Frome, Dorset. *Journal of Animal Ecology*, **51**, 81–96.

May, R. M. (1972), Limit cycles in predator–prey communities. *Science*, **177**, 900–902.

May, R. M. (1976), Models for single populations, in *Theoretical ecology: principles and applications* (ed. R. M. May), Saunders, Philadelphia.

Maynard Smith, J. (1968), *Mathematical ideas in biology*, Cambridge University Press, Cambridge.

Maynard Smith, J. and Slatkin, M. (1973), The stability of predator–prey systems. *Ecology*, **54**, 384–391.

Mech, L. D. (1966), The wolves of Isle Royale. Fauna of National Parks of the United States, *Fauna Series*, **7**, 1–210.

Mech, L. D. (1970), *The Wolf*, Natural History Press, Garden City.

Merriam, J. C. and Stock, C. (1932), The felidae of Rancho La Brea. *Carnegie Institute of Washington Publication*, **422**, 1–231.

Mertz, D. B. (1969), Age-distribution and abundance in populations of flour beetles. I. experimental studies. *Ecological Monographs*, **39**, 1–31.

Mertz, D. B. (1970), Notes on methods used in life-history studies, in *Readings in Ecology and Ecological Genetics* (eds. J. H. Connell, D. B. Mertz and W. W. Murdoch), Harper and Row, New York.

Mertz, D. B. and Wade, M. J. (1976), The prudent prey and prudent predator. *The American Naturalist*, **110**, 489–496.

Michod, R. E. (1979), Evolution of life histories in response to age-specific mortality. *The American Naturalist*, **113**, 531–550.

Milinski, M. (1977a), Experiments on the selection by predators against spatial oddity of their prey. *Zeitschrift fur Tierpsychologie*, **43**, 311–325.

Milinski, M. (1977b), Do all members of a swarm suffer the same predation? *Zeitschrift fur Tierpsychologie*, **45**, 373–388.

Mill, J. S. (1874), *Three Essays on Religion: Nature, the Utility of Religion, and Theism*. Longmans, Green, Reader and Dyer, London.

Miller, G. J. (1969), A new hypothesis to explain the method of food ingestion by *Smilodon californicus* Bovard. *Tebriva*, **12**, 9–19.

Monod, J. (1942), *Recherches sur la croissance des cultures bacteriennes*, Hermann et Cie, Paris.

Mori, H. and Chant, D. A. (1966), The influence of prey density, relative humidity, and starvation on the predacious behavior of *Phytoseiulus persimilis* Athias-henriot (Acarina: Phytoseiidae). *Canadian Journal of Zoology*, **44**, 483–491.

Morris, R. F., Cheshire, W. F., Miller, C. A. and Mott, D. G. (1958), The numerical response of avian and mammalian predators during a gradation of the spruce budworm. *Ecology*, **39**, 487–494.

Murdoch, W. W. (1969), Switching in general predators: experiments on predator specificity and stability of prey populations. *Ecological Monographs*, **39**, 335–354.

Murdoch, W. W. (1970), Population regulation and population inertia. *Ecology*, **51**, 497–502.

Murdoch, W. W. (1971), The developmental response of predators to changes in prey density. *Ecology*, **52**, 132–137.

Murdoch, W. W. and Oaten, A. (1975), Predation and population stability. *Advances in Ecological Research*, **9**, 2–132.

Murie, A. (1944), The wolves of Mount McKinley. Fauna of the National Parks of the United States, *Fauna Series*, **5**, pp. 1–238.

Neill, W. E. (1975), Experimental studies of microcrustacean competition, community composition and efficiency of resource utilization. *Ecology*, **56**, 809–826.

Nellis, C. H., Wetmore, S. P. and Keith, L. B. (1972), Lynx–prey interactions in central Alberta. *Journal of Wildlife Management*, **36**, 320–329.

Nelmes, A. J. (1974), Evaluation of the feeding behaviour of *Prionchulus punctatus* (Cobb), a nematode predator. *Journal of Animal Ecology*, **43**, 553–566.

New York State Conservation Department (1951), *A study of fox control as a means of increasing pheasant abundance*, New York State Conservation Department, Division of Fish and Game, Research Series 3.

Nicholson, A. J. (1957), Comments on the paper of T. B. Reynoldson. *Cold Spring Harbor Symposium in Quantitative Biology*, **22**, 326.

Nicholson, A. J. and Bailey, V. P. (1935), The balance of animal populations. *Proceedings of the Royal Society of London*, **3**, 551–598.

Nisbet, R. M. and Gurney, W. S. C. (1982), *Modeling Fluctuating Populations*, J. Wiley and Sons, New York.

Norton-Griffiths, M. (1967), *A study of the feeding behaviour of the oystercatcher Haematopus ostralegus*. D.Phil. Thesis, Oxford University.

Oaten, A. and Murdoch, W. W. (1977), More on functional response and stability (reply to Levin). *The American Naturalist*, **111**, 383–386.

Orians, G. H. (1969), Age and hunting success in the brown pelican (*Pelicanus occidentalis*). *Animal Behaviour*, **17**, 316–319.

Oster, G. and Takahashi, Y. (1974), Models for age-specific interactions in a periodic environment. *Ecological Monographs*, **44**, 483–501.

Page, G. and Whitacre, D. F. (1975), Raptor predation on wintering shorebirds. *Condor*, **77**, 73–83.

Paine, R. T. (1980), Food webs: linkage, interaction strength and community infrastructure. *Journal of Animal Ecology*, **49**, 667–686.

Paloheimo, J. E. (1971a), A stochastic theory of search: implication for predator–prey situation. *Mathematical Biosciences*, **12**, 105–132.

Paloheimo, J. E. (1971b), On a theory of search. *Biometrika*, **58**, 61–75.

Paloheimo, J. E. (1979), Indices of food preference by a predator. *Journal of the Fisheries Research Board of Canada*, **36**, 470–473.

Partridge, B. L. (1980), The three dimensional structure of fish schools. *Behavioral Ecology and Sociobiology*, **6**, 277–288.

Partridge, B. L. and Pitcher, T. J. (1979), Evidence against a hydrodynamic function for fish schools. *Nature*, **279**, 418–419.

Pastorok, R. A. (1981), Prey vulnerability and size selection by *Chaoborus* larvae. *Ecology*, **62**, 1311–1325.

Paynter, M. J. B. and Bungay, H. R. III (1969), Dynamics of coliphage infections, in *Fermentation Advances* (ed. D. L. Perlman), Academic Press, New York.

Pearson, O. P. (1966), The prey of carnivores during one cycle of mouse abundance. *Journal of Animal Ecology*, **35**, 217–233.

Pearson, O. P. (1971), Additional measurements of the impact of carnivores on California voles (*Microtus californicus*). *Journal of Mammalogy*, **52**, 41–49.

Peterson, R. O. (1976), *The role of wolf predation in a moose population decline*, Proceedings of the First Conference on Scientific Research in the National Parks, New Orleans.

Peterson, R. O. (1979), The wolves of Isle Royale – new developments, in *The Behavior and Ecology of Wolves* (ed. E. Klinghammer), pp. 3–18, Garland STPM Press, New York.

Peterson, R. O. and Allen, D. L. (1974), Snow conditions as a parameter in moose–wolf relationships. *Naturaliste Canadien*, **101**, 481–492.

Peterson, R. O. and Page, P. (1984), Wolf–moose fluctuations at Isle Royale National Park, Michigan, USA. *Acta Zoologica Fennica* (in press).

Pimentel, D. (1961), Animal population regulation by the genetic feed-back mechanism. *The American Naturalist*, **95**, 65–79.

Pimentel, D. (1968), Population regulation and genetic feedback. *Science*, **159**, 1432–1437.

Pimentel, D. and Al-Hafidh, R. (1965), The coexistence of insect parasites and hosts in laboratory populations. *Annals of the Entomological Society of America*, **56**, 676–678.

Pimentel, D., Nagel, W. P. and Madden, J. L. (1963), Space–time structure of the environment and the survival of parasite–host systems. *The American Naturalist*, **97**, 141–168.

Pimentel, D. and Stone, F. A. (1968), Evolution and population ecology of parasite–host systems. *Canadian Entomologist*, **100**, 655–662.

Pimentel, D., Levin, S. A. and Olson, D. (1978), Coevolution and the stability of exploiter–victim systems. *The American Naturalist*, **112**, 119–125.

Pimm, S. L. (1982), *Food Webs*, Chapman and Hall, London.

Popper, K. R. (1959), *The Logic of Scientific Discovery*, Hutchinson and Co., London.

Potts, G. R. (1980), The effects of modern agriculture, nest predation and game management on the population ecology of partridges (*Perdix perdix* and *Alectoris rufa*). *Advances in Ecological Research*, **11**, 2–81.

Powell, G. V. N. (1974), Experimental analysis of the social value of flocking by starlings (*Sturnus vulgaris*) in relation to predation and foraging. *Animal Behaviour*, **22**, 501–505.

Rashevsky, N. (1959), Some remarks on the mathematical theory of nutrition of fishes. *Bulletin of Mathematical Biophysics*, **21**, 161–183.

Recher, H. F. and Recher, J. A. (1969), Comparative foraging efficiency of adult and immature little blue herons (*Florida caerula*). *Animal Behaviour*, **17**, 320–322.

Ricklefs, R. E. (1973), *Ecology*, Chiron Press, Newton.

Ricklefs, R. E. (1979), *Ecology*, 2nd edn, Chiron Press, New York.

Robeson, S. B., Crissey, W. F. and Darrow, R. W. (1949), *A study of predator control on Valcour Island*, New York State Conservation Department Research Series 1.

Rodgers, W. A. (1977), Seasonal changes in group size amongst five wild herbivore species. *East African Wildlife Journal*, **15**, 175–190.

Rosenzweig, M. L. (1969), Why the prey curve has a hump. *The American Naturalist*, **103**, 81–87.

Rosenzweig, M. L. (1971), Paradox of enrichment: destabilization in ecological time. *Science*, **171**, 385–387.

Rosenzweig, M. L. (1973a), Exploitation in three trophic levels. *The American Naturalist*, **107**, 275–294.

Rosenzweig, M. L. (1973b), Evolution of the predator isocline. *Evolution*, **27**, 84–94.

Rosenzweig, M. L. (1978), Aspects of biological exploitation. *The Quarterly Review of Biology*, **52**, 371–380.

Rosenzweig, M. L. and MacArthur, R. H. (1963), Graphical representation and stability conditions of predator–prey interactions. *The American Naturalist*, **47**, 209–223.

Roughgarden, J. (1979), *Theory of Population Genetics and Evolutionary Ecology: An Introduction*, Macmillan, New York.

Royama, T. (1970), Factors governing the hunting behaviour and selection of food by the great tit (*Parus major* L.). *Journal of Animal Ecology*, **39**, 619–668.

Royama, T. (1971), A comparative study of models of predation and parasitism. *Researches on Population Ecology* Supplement, **1**, 1–91.

Rudebeck, G. (1950), The choice of prey and modes of hunting of predatory birds with special reference to their selective effect. *Oikos*, **2**, 65–88.

Rudebeck, G. (1951), The choice of prey and modes of hunting of predatory birds with special reference to their selective effect. *Oikos*, **3**, 200–231.

Rusch, D. H., Meslow, E. C., Doerr, P. D. and Keith, L. B. (1972), Response of great horned owl populations to changing prey densities. *Journal of Wildlife Management*, **36**, 282–296.

Salt, G. W. (1967), Predation in an experimental protozoan population (*Woodruffia-Paramecium*). *Ecological Monographs*, **37**, 113–144.

Salt, G. W. (1974), Predator and prey densities as controls of the rate of capture by the predator *Didinium nasutum*. *Ecology*, **55**, 434–439.

Salt, G. W. (1979), Density, starvation, and swimming rate in *Didinium* populations. *The American Naturalist*, **113**, 135–143.

Salt, G. W. and Willard, D. E. (1971), The hunting behavior and success of Forster's tern. *Ecology*, **52**, 989–999.

Sarason, S. B. (1977), *Work, aging, and social change*, The Free Press, New York.

Schaffer, W. M. (1981), Ecological abstraction: the consequences of reduced dimensionality in ecological models. *Ecological Monographs*, **51**, 383–401.

Schaffer, W. M. and Rosenzweig, M. L. (1978), Homage to the red queen. I. coevolution of predators and their victims. *Theoretical Population Biology*, **14**, 135–157.

Schnell, J. H. (1968), The limiting effects of natural predation on experimental cotton rat populations. *Journal of Wildlife Management*, **32**, 698–711.

Seghers, B. H. (1974), Schooling behavior in the guppy (*Poecilia reticulata*): an evolutionary response to predation. *Evolution*, **28**, 486–489.

Shettleworth, S. J. (1972), Constraints on learning, in *Advances in the Study of Behavior* (eds D. S. Lehrman, R. A. Hinde and E. Shaw), Vol. 4, Academic Press, New York.

Siegfried, W. R. and Underhill, L. G. (1975), Flocking as an anti-predator strategy in doves. *Animal Behaviour*, **23**, 504–508.

Simberloff, O. (1983), Competition theory, hypothesis testing, and other community ecological buzzwords. *The American Naturalist*, **122**, 626–635.

Simon, H. (1969), *The Sciences of the Artificial*, MIT Press, Cambridge, Massachusetts.

Simpson, G. G. (1941), The function of saber-like canines in carnivorous mammals. *American Museum Novitates*, **1130**, 1–12.

Sjöberg, S. (1980), Zooplankton feeding and queueing theory. *Ecological Modelling*, **10**, 215–225.

Slade, N. A. and Balph, D. F. (1974), Population ecology of Uinta ground squirrels. *Ecology*, **55**, 989–1003.

Slobodkin, L. B. (1961), *Growth and Regulation of Animal Populations*, Holt, Rinehart and Winston, New York.

Slobodkin, L. B. (1964), Experimental populations of Hydrida. *Journal of Animal Ecology*, **33**, 131–148.

Slobodkin, L. B. (1974), Prudent predation does not require group selection. *The American Naturalist*, **108**, 665–678.

Smith, J. N. M. (1974), The food searching behaviour of two European thrushes. I. Description and analysis of search paths. *Behaviour*, **48**, 276–302.

Smith, J. N. M. (1975), The food searching behaviour of two European thrushes. II: The adaptiveness of the search patterns. *Behaviour*, **49**, 1–61.

Smith, R. H. and Mead, R. (1974), Age structure and stability in models of prey–predator systems. *Theoretical Population Biology*, **6**, 308–322.

Solomon, M. E. (1949), The natural control of animal populations. *Journal of Animal Ecology*, **18**, 1–35.

Soper, R. S., Delyser, A. J. and Smith, L. F. R. (1976), The genus *Massospora* entomopathogenic for cicadas. II. biology of *Massospora levispora* and its host *Okanagana rimosa*, with notes on *Massospora cicadina* on the periodical cicadas. *Annals of the Entomological Society of America*, **69**, 89–95.

Sparrowe, R. D. (1972), Prey catching behavior in the sparrow hawk. *Journal of Wildlife Management*, **36**, 297–308.

Starfield, A. M., Smuts, G. L. and Shiell, J. D. (1976), A simple wildebeest population model and its applications. *South African Journal of Wildlife Research*, **6**, 95–98.

Starfield, A. M., Shiell, J. D. and Smuts, G. L. (1981), Simulation of lion control strategies in a large game reserve. *Ecological Modelling*, **13**, 17–28.

Stewart-Oaten, A. (1982), Minimax strategies for a predator–prey game. *Theoretical Population Biology*, **22**, 410–424.

Strauss, R. E. (1979), Reliability estimates for Ivlev's electivity index, the forage ratio, and a proposed linear index of food selection. *Transactions of the American Fisheries Society*, **108**, 344–352.

Taitt, M. J. and Krebs, C. J. (1983), Predation, cover, and food manipulations during a spring decline of *Microtus townsendii*. *Journal of Animal Ecology*, **52**, 837–848.

Takafuji, A. (1977), The effect of the rate of successful dispersal of a phytoseiid mite, *Phytoseiulus persimilis* Athias-Henriot (Acarina: Phytoseiidae) on the persistence in the interactive system between the predator and its prey. *Researches on Population Ecology*, **18**, 210–222.

Tamarin, R. H. (1977), Demography of the beach vole (*Microtus breweri*) and the meadow vole (*Microtus pennsylvanicus*) in southeastern Massachusetts. *Ecology*, **58**, 1310–1321.

Tamarin, R. H. (1978), Dispersal, population regulation and K-selection in field mice. *The American Naturalist*, **112**, 545–555.

Taylor, R. J. (1974), Role of learning in insect parasitism. *Ecological Monographs*, **44**, 89–104.

Taylor, R. J. (1976), Value of clumping to prey and the evolutionary response of ambush predators. *The American Naturalist*, **110**, 13–29.

Taylor, R. J. (1977), The value of clumping to prey: experiments with a mammalian predator. *Oecologia*, **30**, 285–294.

Taylor, R. J. (1979), The value of clumping to prey when detectability increases with group size. *The American Naturalist*, **113**, 299–301.

Taylor, R. J. (1981), Ambush predation as a destabilizing influence upon prey populations. *The American Naturalist*, **118**, 102–109.

Thompson, D. J. (1975), Towards a predator–prey model incorporating age structure: the effects of predator and prey size on the predation of *Daphnia magna* by *Ischnura elegans*. *Journal of Animal Ecology*, **44**, 907–916.

Thompson, W. A., Vertinsky, I. and Krebs, J. R. (1974), The survival value of flocking in birds: a simulation model. *Journal of Animal Ecology*, **43**, 785–820.

Tinbergen, L. (1960), The natural control of insects in pinewoods. I. factors influencing the intensity of predation by songbirds. *Archives Neerlandaises de Zoologie*, **13**, 266–336.

Tinbergen, N., Impekoven, M. and Franck, D. (1967), An experiment on spacing out as a defense against predation. *Behaviour*, **28**, 307–321.

Tostawaryk, W. (1972), The effect of prey defense on the functional response of *Podisus modestus* (Hemiptera: Pentatomidae) to densities of the sawflies *Neodiprion swainei* and *N. pratti banksianae* (Hymenoptera: Neodiprionidae). *Canadian Entomologist*, **104**, 61–70.

Trautman, C. G., Frederickson, L. F. and Carter, A. V. (1973), *Relationship of red foxes and other predators to populations of ring-necked pheasants and other prey, 1964–71*. South Dakota Department of Game, Fish, and Parks, P-R Progress Report, Project W-75-R-9, Job F-8.2-9.

Tsuchiya, H. M., Drake, J. F., Jost, J. L. and Fredrickson, A. G. (1972), Predator–prey interactions of *Dictyostelium discoideum* and *Escherichia coli* in continuous culture. *Journal of Bacteriology*, **110**, 1147–1153.

Utida, S. (1953), Interspecific competition between two species of bean weevil. *Ecology*, **34**, 301–307.

Utida, S. (1957), Population fluctuation, an experimental and theoretical approach. *Cold Spring Harbor Symposium on Quantitative Biology*, **22**, 139–151.

Valerio, C. E. (1975), A unique case of mutualism. *The American Naturalist*, **109**, 235–238.

Vance, R. R. (1978), Predation and resource partitioning in one predator–two prey model communities. *The American Naturalist*, **112**, 797–813.

Vandermeer, J. (1981), *Elementary Mathematical Ecology*, J. Wiley and Sons, Chichester.

Varley, G. C. (1975), Should we control the use of the word control? *Bulletin of the British Ecological Society*, **6**, 7.

Varley, G. C., Gradwell, G. R. and Hassell, M. P. (1973), *Insect Population Ecology*, Blackwell Publications, Oxford.

Veilleux, B. G. (1979), An analysis of the predatory interaction between *Paramecium* and *Didinium*. *Journal of Animal Ecology*, **48**, 787–803.

Virnstein, R. W. (1977), The importance of predation by crabs and fishes on benthic infauna in Chesapeake Bay. *Ecology*, **58**, 1199–1217.

Volterra, V. (1928), Variations and fluctuations of the number of individuals in animal species living together. *Journal du Conseil international pour l'Exploration de la Mer*, **3**, 3–51.

Wagner, F. H. (1969), Ecosystem concepts in fish and game management, in *The Ecosystem Concept in Natural Resource Management* (ed. G. M. Van Dyne), Academic Press, New York.

Wagner, F. H. and Stoddart, L. C. (1972), Influence of coyote predation on black-tailed jackrabbit populations in Utah. *Journal of Wildlife Management*, **36**, 329–342.

Wangersky, P. J. and Cunningham, W. J. (1957), Time lag in prey–predator population models. *Ecology*, **38**, 136–139.

Wasserzug, R. J. and Sperry, D. G. (1977), The relationship of locomotion to differential predation on *Pseudacris triseriata* (Anura: Hylidae). *Ecology*, **58**, 830–839.

Watson, A. (1970), Key factor analysis, density dependence and population limitation in red grouse, in *Proceedings of the Advanced Study Institute on the Dynamics of Numbers in Populations* (eds M. H. den Boer and G. R. Gradwell), Oosterbeek.

Weihs, D. (1973), Hydromechanics of fish schooling. *Nature*, **241**, 290–291.

Weihs, D. (1975), Some hydrodynamical aspects of fish schooling, in *Swimming and Flying in Nature* (ed. T. Wu *et al.*), Vol. 2, Plenum Press, New York.

White, E. G. and Huffaker, C. B. (1969a), Regulatory processes and population cyclicity in laboratory populations of *Anagasta kühniella* (Zeller) (Lepidoptera: Phycitidae). I. competition for food and predation. *Researches on Population Ecology*, **11**, 57–83.

White, E. G. and Huffaker, C. B. (1969b), Regulatory processes and population cyclicity in laboratory populations of *Anagasta kühniella* (Zeller) (Lepidoptera: Phycitidae). II. parasitism, predation, competition, and protective cover. *Researches on Population Ecology*, **11**, 150–185.

White, J., Lloyd, M. and Zar, J. H. (1979), Faulty eclosion in crowded suburban periodical cicadas: populations out of control. *Ecology*, **60**, 305–315.

Wilbur, H. M., Morin, P. J. and Harris, R. N. (1983), Salamander predation and the structure of experimental communities: anuran responses. *Ecology*, **64**, 1423–1429.

Williams, G. C. (1966), *Adaptation and Natural Selection*, Princeton University Press, Princeton.

Wolfe, M. L. and Allen, D. L. (1973), Continued studies of the status, socialization, and relationships of Isle Royale wolves, 1967–1970. *Journal of Mammalogy*, **54**, 611–635.

Wollkind, D. J. (1976), Exploitation in three trophic levels: an extension allowing intraspecies carnivore interaction. *The American Naturalist*, **110**, 431–447.

Wollkind, D. J., Hastings, A. and Logan, J. (1982), Age structure in predator–prey systems. II. functional response and stability and the paradox of enrichment. *Theoretical Population Biology*, **21**, 57–68.

Wratten, S. D. (1973), The effectiveness of the coccinellid beetle, *Adalia bipunctata* (L.) as a predator of the lime aphid, *Eucallipterus tiliae* L. *Journal of Animal Ecology*, **42**, 785–802.

Wynne-Edwards, V. C. (1962), *Animal Dispersion in Relation to Social Behaviour*, Oliver and Boyd, Edinburgh.

Zareh, N., Westoby, M. and Pimentel, D. (1980), Evolution in a laboratory host–parasitoid system and its effect on population kinetics. *Canadian Entomologist*, **112**, 1049–1060.

Zaret, T. M. (1980), *Predation and Freshwater Communities*, Yale University Press, New Haven.

Zicarelli, J. (1975), *Mathematical analysis of a population model with several predators on a single prey*. Ph.D. Dissertation, University of Minnesota.

Index